6.95

P9-EDY-239

95

COSMOLOGY
NOW

COSMOLOGY
NOW

edited
by
Laurie
John

British Broadcasting Corporation

Published by the
British Broadcasting Corporation
35 Marylebone High Street
London W1M 4AA

ISBN: 0 563 12370 2
First published 1973
Reprinted 1974

Printed in Great Britain by
The Broadwater Press Ltd
Welwyn Garden City, Herts

The plates which appear between pages 48 and 49 are
printed with the kind permission of:
Hale Observatories, plates 1, 2, 3, 4, 5, 6, 7, 8;
Basil Observatory, plate 3; Lick Observatory, plate 4.

Diagrams by Richard Natkiel.

Dennis Sciama is also the author of *Modern Cosmology* and
The Physical Foundations of General Relativity.

CONTENTS

It was my good fortune to be able to read these chapters sequentially when the original lectures were already in process of transformation to the book. At the end I was left with several outstanding impressions. For the first time we have a collection of the views mainly of the younger generation of cosmologists, who are writing in a language which would seem meaningless to an earlier generation—neutron stars, pulsars, and black holes are concepts of our own age. Their exposition is everywhere brilliantly clear and exciting. The excitement is not bred by any dogmatic assertion that a final solution has been reached to the cosmological problem. On the contrary, there is the acceptance that the present ideas may soon be superseded by others which are scarcely apprehended at this moment. In some places, too, the extraordinary thought begins to emerge that the concepts of physical science as we appreciate them today in all their complexity may be quite inadequate to provide a scientific description of the ultimate state of the universe.

A most notable feature of this book is that, apart from one chapter, by Sir Martin Ryle, all the authors are theorists. This bias is merely indicative of the extreme difficulty we find today in making observations which are decisive contributions to the cosmological problem. In these pages, we find, again and again, that the theorists grasp on fragments of new, and often unconfirmed, observational facts and either despair at the difficulty of interpreting them in an otherwise plausible theory or rejoice that they seem to confirm a particular concept. For example, in the first category we see the difficulty which Professor Penrose has in embracing the gravitational waves, believed to have been discovered by Weber, emanating from the centre of the galaxy. The concept of the black hole, so helpful in the explanation of other modern observations, seems to fail as an explanation of these observations which apparently imply that the energy carried away from the galactic centre by the gravitational waves is equivalent to several thousands of solar masses per year. Already there is substantial evidence that Weber's results are erroneous, at least to an important degree, and so one of the difficulties foreseen by Penrose probably does not exist.

In the second category we find that the surprising discovery of the microwave background radiation by Penzias and Wilson in 1965 leads to a confident feeling by several of the authors that this is, indeed, the relic radiation from the initial hot big bang in which the universe originated over 10,000 million years ago. The reader may feel that if only Dr Sciama's point could be met—that a measurement of the intensity of the radiation at a wavelength of less than a milli-

Introduction
by
Sir Bernard
Lovell
OBE, FRS

Director of
Nuffield Radio
Astronomy
Laboratories

metre would show decisively whether or not the radiation was black body radiation at a temperature of 2·7 degrees absolute—then there could be little further room for argument that the cosmological problem had reached a nearly decisive solution. Decisive, that is, in the sense that the steady state and other evolutionary models of the universe could be finally disregarded. Well, this measurement has now been made using equipment carried in a rocket. The microwave background radiation seems to have been measured satisfactorily down to a wavelength of 0·4 millimetres. These observations were made early in 1973 by a group of scientists from the Los Alamos laboratories of the University of California.

Combined with the earlier measurements at longer wavelengths these new results seem to prove decisively that the intensity of this background radiation increases as the wavelength is decreased, reaches a peak at a wavelength of 1 millimetre, and then falls off. It is exactly compatible with the concept that the measurements refer to an isotropically distributed, 2·7 degree absolute, background radiation. A reading of this book will show how fashionable the interpretation of this in terms of the hot big bang universe has become.

The theory of the steady state universe seems to be in retreat. If the microwave background radiation has the cosmological significance attached to it by several of the authors in this book then the indication of a singular condition of the universe in a past epoch seems inevitable and this is entirely incompatible with the steady state. Apart from the existence of the background radiation there are other observations which encourage this opinion that the universe is in an evolutionary state from a past condition of high density. However, no one acquainted with the contortions of theoretical astrophysicists in the attempt to interpret the successive observations of the past few decades would exhibit great confidence that the solution in favour of the hot big bang would be the final pronouncement in cosmology.

What observations of the universe now would be most likely to place restrictions on possible theoretical interpretations in cosmology? The answer is probably the measurement of the deceleration parameter and this requires the extension of the Hubble red-shift—apparent magnitude line to greater distances. Most cosmologists would agree that if this relation could be established unambiguously so that the curvature of the line could be determined then a decisive indication of the past and future condition of the universe would emerge. General relativity theory—and in a restrictive sense, Newtonian cosmology—establish unambiguous relations between the deceleration parameter and the possible states of the universe. The great difficulty is that the slope of this Hubble line is nearly identical for all possible universes, including the steady state, out to red-shifts of nearly half the velocity of light. It is for greater red-shifts that the significant differences in the prediction of the theory emerge. Since 1951, when the discovery of the radio galaxies encouraged the hope of greater penetration into the past history of the universe, many astronomers have concentrated their observations on this problem. When quasars were discovered it seemed that the means were to hand, at last, for the extension of the Hubble line to the really decisive past epochs of the universe. However, as described in these chapters, the result has been disappointing and inconclusive for two reasons. First, the points for the quasars seem to produce merely a scatter diagram embracing a wide range of possibilities for the extension of the Hubble line to greater red-shifts. Secondly, violent arguments are at present in progress as to whether the measured red-shifts for the quasars really are wholly a cosmological effect. A dozen years after the discovery of the quasars, as far as the measurement of the deceleration parameter from the red-shift—magnitude diagram is concerned, all possible cosmologies remain plausible.

However, this confused situation may not remain for long. Since these chapters were compiled red-shifts have been measured exceeding a value of about 2·9. In the first four months of 1973, two quasars with red-shifts of about 3·5 were identified. The important point about these measurements was not only that an apparent barrier at a certain lookback time into the universe had been surmounted but also that the two quasars in question were neutral in colour, not sharing the blue excess of the quasars identified at smaller red-shifts. This immediately stimulated the publication of a list of 24 hitherto unidentified objects which emitted radio-waves, whose positions were known with an accuracy sufficient to identify them unambiguously with similar objects of neutral colour on the Palomar sky atlas. The investigation of the spectra of these quasars is confidently expected to make possible an even further extension of the Hubble line. Of course, these may show the same scatter as those of smaller red-shifts, but again, recent investigations of the differing types of radio spectra of the quasars show most hopeful signs of yielding more homogeneous groupings.

If these measurements and the derivatives which are possible—such as the variation of angular diameter as a function of red-shift—indicate a value for the deceleration parameter appropriate to an expansion of the universe from a condition of high density some 10,000 million to 15,000 million years ago, and if no alternative explanation of the microwave background radiation emerges, then cosmology will have reached a decisive stage. The conclusion would then seem inevitable that the universe of our present-day observation began in the high-temperature, high-density condition so vividly described in these chapters. But at the present moment these issues remain to be settled. On the whole, the discoveries of the last twenty years have served to uncover more problems about the universe than have been solved. The investigation of the universe in the X-ray region of the spectrum by apparatus carried in earth satellites has only recently begun and the study of gamma rays from space has scarcely commenced. It would be wise to expect a revolution in our knowledge of the universe from these observations at least as great as that following the discovery of the radio waves from space. *Cosmology Now* should be read as the authors intended—the best description of the universe which can be given today—and not as a final solution to a problem whose immensity we still may not be able even to envisage.

Cosmology faces scientists with a challenge that, for sheer magnitude, cannot be surpassed. Its field of inquiry is the entire universe. The challenge: to describe that universe as it was, is now and will be in the far future, and to relate this description to our earthly experience. Little wonder that for centuries this enormous task has been the province of the theologian. Our attempts to use the language of science to describe the startling information now being gathered from the depths of space is straining our understanding to the uttermost. Even, some would say, beyond. So the stark question to be faced at the outset is whether cosmology is a proper field for the scientist at all.

Professor Sir Hermann Bondi is now chief scientific adviser to the Ministry of Defence, but here he writes as one who has contributed a wealth of ideas to what he would unhesitatingly describe as 'the science of cosmology'. L. H. J.

SETTING
THE
SCENE
Professor
Sir Hermann
Bondi
KCB, FRS, FRAS

Cosmology is the field of thought that deals with the structure and history of the universe as a whole. Human curiosity has been concerned with what the universe is like and how it began for as long as recorded history. There have been many speculations, philosophies and theologies concerned with it. Nowadays we regard cosmology as a branch of science, or to be more precise, a branch of astronomy. If we say that something is a branch of science, we have got to be pretty clear about what we mean by 'a science'. To define a science is the province of the philosopher of science, and naturally there are various schools of thought on this subject. Personally I am a follower of Karl Popper, who in my view has described the scientific process with tremendous accuracy, veracity and clarity. Let me give a very abbreviated version of Popper's Theory of Demarcation, that is the demarcation between science and other human pursuits. According to him, it is the task of the scientist, guided by the knowledge of his time, to propose a theory that takes into account what is known, but which, over and above this, forecasts what future experiments and observations should show. It is only if a theory submits itself to empirical tests that one can call it scientific. If such an empirical test goes against the theory, then the theory has been disproved. If it agrees with the forecasts of the theory, then it becomes the task of the theorist to go on making more and more forecasts, to go on sticking his neck out. A theory is scientific only as long as it lives dangerously. If it is not at risk, it is not part of science.

According to this view, science doesn't prove anything at all, but it disproves an awful lot. Why should there be this

Chapter 1

emphasis on disproof? The answer lies essentially in the philosophical theory of induction. The evidence we gather always deals with *particular* experiments, *particular* observations and situations, whereas a scientific theory is a *general* statement. How many apples do I have to observe to fall after I have released them before I can state that all apples when released will fall? Quite clearly there is no limit to that number. I can never make a logical deduction from having observed a finite number of apples that all apples will behave like this. But if I state theoretically that all apples will behave like this, then if a single apple rises, the theory is disproved. In this way we get a very clear view of what a scientific theory should be and what it must do. Above all it stresses very much the highly imaginative character of science. The essential thing in science is for the scientist to think up a theory. There is no way of mechanising this process; there is no way of breaking it down into a science factory. It always requires human imagination, and indeed in science we pay the highest respect to creativity, to originality. It is, of course, clear that since every theory must live dangerously, the casualty rate is pretty high. So we do not honour scientists for being right; it is never given to anybody to be always right. We honour scientists for being original, for being stimulating, for having started a whole line of work. Science is the most human of endeavours because it depends on co-operation, it depends on people testing each other's work and it depends on people taking notice of each other.

Given this definition of science, how can one regard the subject of the structure of the universe as a whole as a fit subject for the scientific approach? Clearly, this can be done only if it is reasonable to make theories about the structure and history of the universe that can be tested by experiment and observation; and this indeed has been the history of the subject. We have had many theories of cosmology that originally appeared to be very sound, but which have gradually been disproved through experiment or observation. As the growth of technology allows us to build increasingly better experimental equipment the tests become more and more severe. What may have passed a test at one moment in time cannot pass some time later.

It is obvious that in cosmology one is dealing with a vast subject, and in order to be able to tackle it at all one has to make certain assumptions. Now a working hypothesis—one that is accepted by almost every scientist working in the field—is the notion that the universe viewed on a sufficiently large scale is uniform; that it is the same elsewhere as it is in our astronomical neighbourhood. ('Neighbourhood' here must mean a pretty large region.) This does not necessarily mean that it is the same at all times. It may be changing in time. Uniformity does not necessarily imply that it is a motionless universe. But the only type of motion compatible with uniformity (barring some oddities that we won't be concerned with) is what we call a uniform expansion or contraction in which the relative velocity of any two points is always along the line joining them, and is, roughly speaking, proportional to the distance between them. If you want to make a simple image of this for yourself, just imagine, as a one-dimensional model, a long row of objects (see Fig. 1). They are all at an equal distance from each other. If we imagine that we stand on one of them and that our neighbours either side recede from us at 1 km per second, the ones next to our neighbours at 2 km per second, the ones one step further away at 3 km per second, and so on, then we have such a uniform system. For our neighbour will see both his neighbours (of which we are one) receding at 1 km per second; his neighbours next but one at 2 km per second; and so on. The system is uniform in the sense that each one of the observers in it gets the same picture of it as any other does.

I've been giving a one-dimensional example. You can, of course, equally think in terms of three-dimensional examples. Imagine a large cube of rubber. At the word 'go' people on each side of it start squeezing it until it is half its size. Then

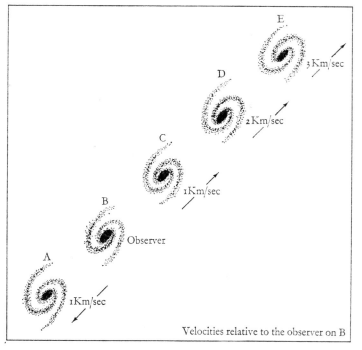

Velocities relative to the observer on B

Fig. 1: A uniform universe. No other model meets the conditions of uniformity and dynamic motion.

if you think of any two particles of that block of rubber after the compression has been applied, wherever these particles might have been originally, their distance will have been halved and the orientation of one vis-à-vis the other will have been unchanged. So again you have motion along the line joining the particles with the velocity proportional to the distance between them, because if they were far from each other they have had to travel a long way to halve this distance during the period of compression; if they were close to each other they've had to travel only a very small way. This, then, is the only kinematically possible picture of a uniform system remaining uniform.

So far I have only talked about the geometry of a uniform universe in the sense of a situation that looks uniform to everybody involved in it. If next we try to see how the forces, and particularly the gravitational forces, in a large system would affect the situation, it is possible to use various theories of dynamics. We can use the slightly obsolete dynamics of Newton, which we know is rather good when relative velocities are small, but begins to be a little shaky when you get up towards the velocity of light. Or we can take the more complicated, but more comprehensive relativistic dynamics which copes with high velocities just as well as with low ones. Whichever assumption you start with, you find that the natural, the obvious, the most likely, perhaps the only possible state of a uniform system is not to be at rest but to be in motion—in motion of exactly the type I've been discussing, in which the relative velocity of two particles is along the line joining them and proportional to the distance between them.

As a matter of fact this point was first discovered by scientists late in the 19th century who found that Newtonian dynamics naturally led to a moving universe, but the prejudices in favour of a static system were so tremendous that their work was not taken any notice of and forgotten. The next person on the scene was Einstein. He too was much influenced by the notion of a static universe and he forced his equations into a form that allowed them to have such a static solution. Moreover, he thought that this solution was the only solution of these equations. Indeed it came as a surprise, intended as a mathematical curiosity only, when the Dutch astronomer De Sitter shortly afterwards showed that Einstein's equations also had a solution that gave a universe in motion, i.e. a universe with a motion of expansion, velocity proportional to distance. Although this was originally conceived of as a purely mathematical construction, in a few years observational evidence led people to regard this model as the best mathematical description of what the universe was like in practice.

The first empirical evidence came from the astronomer Slipher, who measured the spectra of the galaxies, and found a tremendous predominance of red-shifts as opposed to blue-shifts. It is a well-known result from theory, tested by countless experiments, that the colour of the light from a receding source is shifted towards the red—a red-shift. The frequencies of the spectral lines characteristic of the source are all lowered in the same proportion. Correspondingly if the source is approaching there is a blue-shift in which the frequencies of the spectral lines are raised. This suggested to Slipher that perhaps there was a motion of expansion going on in the universe. A number of other astronomers took up the challenge under the leadership of Hubble and before the 1920s were out, Hubble had quite clearly established the dominance of red-shifts and, moreover, he had found that the magnitude of the red-shift of the galaxies was closely correlated to how bright they looked to us. The fainter they looked, the greater the red-shift was.

In astronomy, it is no easy task to measure the distance of an object; and indeed in many cases our only indication of the distance of an object like a galaxy or a quasar is how faint or bright it looks. If then, the apparent faintness of Hubble's galaxies is interpreted as a result of their distance, and the red-shift is interpreted as their velocity away from us, then indeed it turns out that the velocity is proportional to the distance. Hubble's results have been extended enormously since he arrived at them, and the velocity-distance relation still holds. Thus both observation and theory lead to a universe in motion rather than a static universe. This result is now very well established.

But it is important that one should be clear about precisely what it is that is established. It is this close correlation of the magnitude of the red-shift and the faintness of the object; and it is most fruitful to assume that the best method of accounting for the red-shift is a recession. When we make this identification we support a great deal more. If the only type of motion compatible with the assumption of uniformity is a recession with the velocity proportional to distance, then this lends very powerful support indeed to our initial assumption of uniformity.

Moreover, there are rather direct tests of uniformity. Due to the difficulty of measuring cosmological distances, it is hard to say very much about uniformity in depth. But at least we know that a uniform system should look much the same in whichever direction you are looking. It should look isotropic; and indeed, this is what we observe, provided we look far enough. Of course, our astronomical neighbourhood is not isotropic. We know on the very small scale of the earth that there is solid earth and ocean and atmosphere. On the larger scale, that there are stars and spaces between the stars. On a yet larger scale, that there are galaxies and spaces between the galaxies. Indeed, our own galaxy is clearly not isotropic because in certain directions you see this bright band of stars across the sky and in others, you do not. But our own galaxy is a local phenomenon. The dimensions of our local galaxy are measured in a few tens of thousands of light years, where a light year is the distance travelled by light in one year—about 6 billion miles. The distances between galaxies are much larger, perhaps of the order of a million light years. Even then, on this scale, you still get distinct non-uniformities due to galaxies being grouped into clusters, ours being a member of the local cluster. But the more you go to large distances, the greater your statistical sample, the more all these non-uniformities disappear until, with very faint objects, you have a very good measure of isotropy indeed. So, the expanding universe is the background into which all current theories and probably all theories for many years to come will have to fit.

The indications that our universe is expanding leads naturally to the idea that its history and geography are linked. Light has a finite velocity, however large it may seem by ordinary terrestrial standards. Virtually all our knowledge of the universe is obtained by observing radiations of various kinds (visible light, radio waves, ultra violet rays, X-rays, gamma rays) all travelling in straight lines at the speed of light, the only real exception being local knowledge of the past (the rocks of the earth, of the moon, and perhaps other planets). Thus we see *now* light emitted by objects 10 million years ago at a distance of 10 million light years, and equally we *now* receive radiation emitted 1,000 million years ago by objects 1,000 million light years away, and so on. The more distant the objects we study, the longer ago is the state in which we see them. If we want to know what the universe was like long ago, we can do so by putting together the information we obtain from distant objects.

Thus, in a changing universe, we cannot possibly test its uniformity by looking into the depths of space and then comparing our findings with nearer regions. For we see the distant regions as they were when the universe was much younger, the nearer regions as they were when the universe was nearly its present age.

How old is the universe? Much intricate argument has

gone into attempts to answer this question, and some of it will be discussed in later chapters. However, it is easy enough to define a quantity that must in some way be characteristic of the age of the universe, a standard that tells what kind of times (and therefore distances) we should contemplate. I have already referred to the red-shift—faintness relation, interpreted as a proportionality of velocity and distance. Thus, the distance of a remote galaxy, divided by its velocity of recession, yields the *same* quantity, whichever remote galaxy is chosen. This quantity is a time, to wit the time this galaxy would have taken, at its presently observed velocity, to reach its presently estimated distance. The difficult task of estimating distances means that there are considerable uncertainties in our knowledge of this time, but the best current estimate is between 10,000 and 20,000 million years, a few times the age of the oldest rocks of the earth. It is times of this order that we must think about as characteristic of our universe, just as it is distances of 10,000 or 20,000 million light years that are characteristic of the size of our universe.

A peculiar feature of our universe is that it is so transparent. We think that at least some of the objects we can observe, in radio or optical wavelength, are so far away that their radiation has travelled to us over several thousands of millions of light years. If it were not for this transparency, we would know far less than we do about it all. Conversely, the fact that our observations span distances and times that are a sizeable fraction of the characteristic distances and times of the universe gives us considerable confidence in the relevance of these observations to cosmology. Since the universe, thanks to the velocity—distance relation, has characteristic lengths and times, we need not fear that we only look at so puny a fraction of the whole that our knowledge has purely local significance.

Of course, on this scale we are talking about the universe only in a very broad brush sense. We are only talking about overall features, we are not talking about what it consists of, what the stars are like in the galaxies, what they look like, what the radiations are like at optical wavelength and infrared and radio wavelengths. A vast body of knowledge has arisen in these fields in the last decade or two through improvements in the means of observation and the later chapters in this book will describe these observations and show how they affect the theoretical picture.

Meanwhile, the first task of cosmology is to make a theoretical model of the universe, a model that in its broadest features must fit in above all with the notions of uniformity and the motion of recession, and that then has to be a home for such developments as give rise to the sources of radiation and the galaxies and the clusters of galaxies that we see.

However, a peculiarity of cosmology is that although we rely so very firmly on this assumption of uniformity on a large scale, nobody has ever given a precise formulation of it. And indeed there are certain difficulties involved. It is a truism that the universe is either infinite or finite. The notion of the finite universe is not as strange now as it once was; it is simply the idea that through a form of space curvature, however straight you like to go, you will eventually come back to where you started from. This is the well-known finite but boundless model of the universe. On the other hand, no one has yet made any observations to show that the universe may not be infinite in extent. In an infinite universe, we could imagine an asymptotic approach to uniformity where we simply say that the larger the volume you consider, the more any average for the volume approaches the same value around whichever point your volume is centred. On the other hand, you can make the criticism that any region we can actually survey astronomically must be finite. This, therefore, is an insignificant portion of the whole infinite universe, and to infer any properties of the universe as a whole—an infinite universe —from our knowledge of such a finite example doesn't make much sense. On the other hand, if the universe were finite, we would also be in difficulties, because then there would be no easy way of describing the approach to uniformity. At any moment, there would be *a* biggest galaxy somewhere in it, *a* brightest star, *a* most massive gas cloud, so uniformity doesn't make all that much sense in a finite universe either.

But could it be that it is just our universe with its motion of recession that permits a sensible notion of uniformity to exist? No precise formulation of this idea has yet been made, but perhaps I can make the point plausible. As I have been saying, when an object recedes from us, the light is shifted towards the red, but this red-shifted light is also weakened; not just by the distance but by the velocity; and at large distances where the velocity is high, this indeed becomes the predominant factor. So if we look out to a certain distance and want to look further, it becomes progressively more difficult, because in addition to the normal weakening of the intensity of light with distance, there is the weakening due to the higher velocity of recession sources further away from us. Thus there is a law of diminishing returns. In our astronomical neighbourhood, if I double the aperture of my telescope I can, broadly speaking, see twice as far. At large distances, this is by no means true and at very large distances doubling your aperture would only allow you to see a few per cent further, according to our present understanding. So we are not running into a brick wall, but we are running into a sort of fog. A fog which never suggests, 'here you stop and you can't take a larger sample than this'; you can, but it gets a lot more difficult. This may be (and I stress that I can only say 'may be') the compromise that will one day allow a sensible formulation of uniformity to be given in spite of the dilemma that I stressed; the difficulty in the infinite case and the difficulty in the finite case.

I should also mention another very remarkable characteristic of our universe, and that is that it is a dark cold place in which there are very bright, hot objects. Now, this is odd because we tend to think of our universe as old, and we know that temperature differences gradually die out. If you put hot coffee and an ice cube into a vacuum flask, after some time it will all be the same temperature, but there is not the slightest approach to such a situation in our universe. There are still vast temperature contrasts; a tremendous thermodynamic disequilibrium, as the physicists would call it. It may well be that the recession is a powerful agent in maintaining such a disequilibrium. It may be that the universe is, in fact, relatively young; but again this disequilibrium, these vast temperature and brightness differences are something for which any sensible theory of the universe must account.

One of the earliest exercises in theoretical cosmology is concerned with precisely these ideas. The argument is generally known as Olbers' paradox, after the astronomer who published it in 1826, although there was an earlier description of it (unknown to Olbers) given by J. P. L. de Cheseaux of Lausanne in 1744.

In these early discussions, it was natural to assume the universe to be both uniform and unchanging, and it was taken for granted that the whole system of 'fixed' stars was truly static. Even today, this set of assumptions would seem the most sensible and natural one that could be made for a start. It is only in the last fifty years that we have learned, as described above, that the universe is not static, and only in the last twenty years that observations have been made that appear to indicate in some manner how the universe may be changing. Equally naturally, then as now, the validity of terrestrially discovered physics was accepted as being universal.

The thought that led these authors to their investigations

was that at great distances there should be vast numbers of stars, which, though each of them would be too faint to be seen individually, in their totality should give us a sort of background glow of the sky. What, they asked, should be the intensity of the glow?

The argument is mildly mathematical. Consider a huge spherical shell (containing numerous stars) with a certain radius, and some thickness which is small compared with its radius. The number of stars in the shell, and so the amount of light sent out by all of them together is, in a uniform universe, proportional to the volume of this shell which, given its thickness, varies like the square of its radius, while the intensity of light we receive from each star is inversely proportional to the square of the radius. Thus the total amount of light we receive from the stars in such a shell is independent of its radius. On the basis of our assumptions, we may add another shell just outside the one we have considered, and one outside that, and so on, *ad infinitum*. Since we receive the same amount of light from each shell, the sum total should be infinite. However, not only do the stars in each shell emit light, they also get in the way of light from yet more distant stars. Taking some average size of star, the fraction of shell area occupied by star images is again independent of shell radius, since both shell area and star number are proportional to the square of the radius. Each shell, therefore, obscures the same fraction of sky. Thus we receive less light from the distant, more heavily obscured, shells than from near ones. Summing the amount of light we receive from stars in the various shells, we get a geometric series converging to a finite limit, a limit evidently closely related to the ratio of the radiating power of a star (its average luminosity) to its shadowing area which gives us effectively a 'background' light of the sky equal to the intensity of light on the surface of an average star. The sun being a reasonably average star, and our distance from the sun being about 200 solar radii the intensity of this background light at night turns out to be around 40,000 times broad daylight!

This astonishing result is known as Olbers' paradox. The flagrant contradiction with observation shows that the universe cannot have all the properties implied by Olbers' assumptions (uniform, unchanging, static).

It has always been a matter of surprise that such a simple argument, combined with an obvious observation, could lead to the disproof of a cosmological theory, viz that the universe is uniform, unchanging and static. It will be realised that this is a beautifully direct example of Popper's criteria for a scientific theory: assumptions of the theory from which observable consequences are drawn, plus contradiction with observation equals disproof of theory!

The importance of the paradox is so great that perhaps it is worth giving another proof. It is easily seen that the apparent surface brightness (that is the intensity per solid angle) of an object is independent of its distance. If we fix the solid angle by looking along a tube (such as a rolled up piece of paper) held close to the eye and observe an illuminated featureless surface (say a white wall) then the apparent brightness of the patch of wall seen does not vary as we change our distance from the wall. The further we are from the wall, the bigger the piece of wall we can see through our tube, but the less light we receive from each sqaure centimetre of wall due to our greater distance. The two effects compensate exactly.

In an 'Olbers' universe', in whatever direction we look, tracing the ray of light away from us, it would eventually hit a piece of stellar surface, a piece limited sideways in general by other stars that are in the way. (Though stars look like pin points, they do of course subtend an angle, however small.) Thus the sky should be a mosaic of myriads of pieces of stellar surface covering it in its entirety. The brightness of each piece is independent of the distance of the star concerned, since surface brightness has this property. We know how bright a good average star, the sun, looks. Thus the whole sky should look as bright as the sun, giving us again a few tens of thousands times sunlight.

Note that this result is not upset by any modern knowledge. We know now, as Cheseaux and Olbers did not, that the stars are not uniformly distributed, but congregate in galaxies, but with a uniform unchanging static distribution of galaxies the Olbers' result stands just as much as with stars. Nor is it upset by our knowledge that, in addition to stars, space also contains some dark clouds of dust and gas. For, although indeed such a cloud can absorb starlight, absorbing it raises its temperature. In an Olbers' universe, with plenty of time available, any such cloud would go on absorbing light and therefore getting hotter until it was glowing so strongly that it emitted as much light as it received, in which case it would evidently not reduce the brightness of the sky.

Thus the observation that it is dark at night clearly disproves the theory that the universe is in fact uniform, unchanging and static. As has been said earlier we have significant indications as to the uniformity of the universe, but the most obvious (indeed the only plausible) explanation of the observed red-shift is non-static, i.e. a universal motion of recession fully compatible with uniformity. (The question of the changing character of the universe is discussed in later chapters, but the type of evolution now most in favour makes no contribution to the resolution of Olbers' paradox.) The red-shift and recession, however, completely account for the darkness of the sky. For a motion of recession, as stated before, not only leads to a red-shift but also reduces the intensity of the light received. Indeed, it is a square law effect, so that if the frequence of the light is reduced by a certain factor the intensity is reduced by that factor squared. Thus we receive less light from distant shells than in a static model, so much less that we get a very dark sky.

Indeed we can get a rough numerical check on this. The transparency of the universe implies that the bulk of Olbers hypothetical flood of light would come from very far away, indeed from absurd distances like a billion billion (10^{24}) light years. But the red-shift means that we receive little from distances of more than 10,000 million (10^{10}) light years. Since each Olbers' shell makes the same contribution, it should therefore be 100 billion (10^{14}) or so times darker than Olbers' paradox suggests, perhaps 2.5 thousand million times darker than broad daylight, or about 25,000 times darker than the light of the full moon. This would be approximately right for an average point in the universe, between the galaxies and thus not illuminated by the light of numerous near stars as we, citizens of a big galaxy, are.

Thus the Olbers' paradox teaches us three things: first that cosmology is a science since a theory can be disproved by empirical results. Secondly that some very simple observations, such as that it is dark at night, may have very significant consequences. And thirdly that the universal motion of the recession of the galaxies is the major feature of our universe.

To put all this in slightly different terms. It is a peculiarity of our universe that it is dark and cold, but contains very hot bright bodies, the stars. It is thus, to use again the physicist's language, in a state of extreme thermodynamic disequilibrium, i.e. it contains very large temperature differences. Any theory of the universe that is not, at the very least, compatible with this property (and preferably accounts for it) is not worth thinking about. Thermodynamic properties tend to be very deep and significant: the fact that our night sky is very black, with very bright points, the stars, in it, may be the profoundest piece of knowledge of the universe that we have.

Having established that the universe is expanding, cosmologists now want to know whether this expansion will go on forever or whether the universe will eventually recontract. One of the most powerful means of finding out involves, in essence, an extension of Hubble's pioneering work on galaxies. The nearest galaxy to our own is the famous nebula in Andromeda, which can be seen with the naked eye. Field glasses will bring perhaps a thousand more to view. But today's powerful telescopes reveal that our galaxy is accompanied in the universe by something of the order of a million million others! A truly astronomical number. A number so huge that it has proved a hopeless task even to catalogue them all.

Dr John Peach, Dean of Brasenose College, Oxford, is among those who are observing this galactic host to gain information which, in many instances, started on its way to us not long after the beginning of time. L. H. J.

The approach to cosmology I am going to discuss might be considered by some as rather old-fashioned. We have grown accustomed during the past few years to the long series of important cosmological discoveries that have been made by exciting new methods which were not available to the pioneers in the subject. In other places in this book you can read about the vitally important information on the large-scale structure of the universe that has been obtained by the methods of radio astronomy, and through microwaves and X-ray astronomy and other developments of the Space Age with its satellites and inter-planetary probes. I am going to concentrate exclusively on what the use of large optical telescopes can tell us about the subject. This is the way that observational cosmology started, and I hope to convince you that there is still a vital contribution to be made by these means. The objects I am going to concentrate on are *galaxies*. It was through a study of these objects sixty years ago that the scientific study of cosmology began and the development of our knowledge of them is the central line of development of the subject.

In the early years of this century galaxies were always referred to as *nebulae*, from the Latin for 'clouds'. On looking at them through a telescope they differ from stars in being fuzzy or cloud-like in appearance. It is amusing to recall that the early cataloguers of nebulae were in many cases much more interested in comets that the nebulae were often being confused with, and the nebulae themselves were regarded as something of a nuisance. On examination in large telescopes some of them show a characteristic spiral structure (plate 1, for example) while others, the elliptical

Chapter 2

galaxies (plate 1), are regular patches of light with a bright centre. Still others, and these are not of direct interest to us as they are very local objects, are highly irregular and are associated physically with hot stars in our own galaxy.

When the first spectra were taken of the spiral and elliptical nebulae it was found that the spectral lines characteristic of stars and interstellar gas appeared, in most cases shifted to longer wavelengths, that is towards the red end of the spectrum. As Hermann Bondi has already explained in Chapter 1, any astronomer seeing this red-shift effect would naturally assume that it reflected the motion of the objects away from the solar system. All stars show a shift of the lines in their spectra either to the red or the blue end of the spectrum, that is, to longer or shorter wavelengths, reflecting their motions with respect to the observer as they move in their paths through our galaxy. But the red-shifts shown by the nebulae were in many cases much larger than stars ever show. It seemed that one was dealing with a type of object quite different from ordinary stars, but until the middle 1920s, although many people guessed what we now know to be their true nature, there was no really compelling explanation of them.

Hubble's demonstration that the nebulae were in fact large numbers of stars at very large distances was probably the most important advance in astronomy since the establishment of the Copernican system, and yet, as with so many important discoveries, on looking back it seems very obvious and unavoidable. Using the 100-inch reflecting telescope of the Mount Wilson Observatory, Hubble took a large number of photographs of one of the brightest of the spiral nebulae, M 31, the Great Nebula in Andromeda (plate 2). He noticed that apparent point sources of light in the spiral arms seemed to change regularly in brightness with time. When he carefully plotted this luminosity variation it was found to resemble closely the variation of brightness characteristic of a class of variable stars called Cepheids, which are common in our own Milky Way. Now it had been established some time previously that the longer the period of variation of these objects, the higher their intrinsic brightness, and a detailed relation had been established enabling one to use the period of a Cepheid and its apparent brightness to find its distance. Hubble boldly applied this relation to the variables he had discovered in Andromeda and found that this implied that M 31 was at a distance much larger even than the distance to the centre of our own galaxy, the Milky Way system. He then proposed a model for the nebulae which is the one that we believe today, namely that they are large aggregations of stars, typically of

around 10,000 million (10^{10}) or 100,000 million (10^{11}) stars
(although this number can vary widely) situated at great
distances from us. We ourselves are situated in a nebula
which we call the galaxy, and looking through the clouds
of stars of our galaxy we see the extra-galactic nebulae, or as
we call them by analogy with our own system, the galaxies.
It is probable that the vast bulk of all the matter in the
universe now exists in the galaxies, which are separated by
almost empty space.

Having found a method of determining distances to
galaxies through observations of their variable stars, Hubble
then turned his attention to the previously unexplained high
velocities of recession of these objects and discovered what is
still the key cosmological fact, namely that the more distant
a galaxy the faster it is receding. When he established this
correlation between velocity and distance it was based on
what are, by the standards of modern extra-galactic
astronomy, very nearby galaxies and as long as the distances
were measured by observation of the variable stars in the
galaxies this was inevitable. Individual stars are only detectable in nearby galaxies; further away they are too faint to be
detected. One has therefore to search for other methods for
measuring the distances of galaxies. One can use the
properties of novae, a type of star that flares up explosively,
becoming intrinsically very bright and therefore visible over
large distances before fading away again. Or one can use the
properties of the large masses of glowing hydrogen gas
embedded in the arms of spiral galaxies which are called
H II regions. Both the novae and the H II regions show a
sufficient uniformity of properties, either in intrinsic brightness or size, to enable one to infer the distance of their
galaxies from their appearance on telescopic photographs.

But there inevitably comes a time when galaxies are so
distant that none of these distance indicators can be properly
resolved in a telescope, and one is forced to base one's
estimate of the distance of the galaxy on the apparent
brightness of the galaxy as a whole. This presents a number
of problems. Not all galaxies are of the same size, type and
intrinsic brightness; there is indeed a variation of more than
a hundred times in the intrinsic brightness of members of
this very differentiated species. One therefore has to have
some way of recognising the type of galaxy before one can
distinguish between two *a priori* equally likely situations, that
is, that one is observing an intrinsically bright object at a
great distance, or an intrinsically faint object at a small
distance. I will return to this point later, but first I would like
to discuss the masses of the galaxies.

How does one even begin to estimate the mass of a galaxy, bearing in mind that we are talking about a faint object which may be many millions of light years away? Well, one can make a first guess by noting that a typical galaxy has the brightness of say 1,000 million (10^9) stars of the type of the sun and if all the stars in the galaxy are similar to the sun then one would expect the mass to be about 1,000 million (10^9) solar masses. It turns out that this is not a bad way of proceeding in the case of the spiral galaxies, because the sun does indeed have about the average mass and average luminosity of a star in one of these galaxies. A more satisfactory approach is to investigate the internal dynamics of the galaxies. We have known since the 1920s that our own galaxy is rotating as a whole in such a way that our sun travels round the centre of the galaxy in a roughly circular orbit once every 200 million years. Similar orbital motions occur in all galaxies and the use of the spectrograph enables us to analyse these motions to give their masses. A simple application of Newtonian dynamics enables us to derive the mass of the earth from the orbital period and orbital radius of an artificial satellite. In a similar way if one can observe an object in orbit around a galaxy and measure its period and orbital radius one can determine the mass of the galaxy. The only objects we can use in this way are pieces of the galaxy itself. If one observes outlying pieces of a galaxy with a spectrograph one finds that in most cases the shift of the spectral lines with respect to the nucleus of the galaxy at its centre indicates that one extreme of the galactic disc has a velocity of approach and the other extreme has a velocity of recession. These velocities can be corrected for the effects of geometrical projection to give us the orbital velocities of the component parts of the galaxy and from these we can infer the mass of the galaxy interior to the points whose velocity we have measured.

This technique has been applied to a number of spiral and elliptical galaxies and we now have a picture of how the masses of galaxies vary with their sizes and types, and, as with their brightness, we find that there is a spread of several factors of ten among the masses of different galaxies. This knowledge together with what we know from the work described above of the distance scale gives us the ability to calculate a very important cosmological parameter: the mean density of matter in our locality. We tot up the masses of all the galaxies in that volume of nearby space for which we have a fairly good idea of their distances and divide the total mass by the volume enclosing it to give a mean density in, say, grams per cubic centimetre. By sampling over successively larger volumes we may arrive at a quantity which may reasonably be called the mean density of matter in the universe as a whole. It turns out to have the unbelievably small value of about 10^{-30} grams per cubic centimetre— some thirty factors of ten smaller than the density of water.

Before we go on to say more about the observations let us think for a few moments about one or two theoretical ideas. You will recall that Hubble established that the further away a galaxy was, the faster it was moving away from us. This observational relationship between velocity and distance is roughly linear. Now we make the simplifying assumption, which underlies the vast majority of cosmological theories, that the universe is roughly the same at every point within itself, implying that there are no privileged positions within it. If we are to accept the assumptions of isotropy and homogeneity, then Hubble's law is the only functional form of the velocity-distance relations allowed.

Now theoretical astronomers have constructed a large number of possible cosmological models, the two main examples of which are described in chapters 5 and 6 of this volume. These models are physically conceivable structures of the material universe, and a complete theory would, hopefully, specify how this structure varies with time. It would say (for example) whether the present observed expansion is going to continue indefinitely into the future, or whether we can expect the expansion to slow down and stop. It should also tell us what has happened in the past. The normal theoretical background to a cosmological model is Einstein's general theory of relativity, which is the currently accepted theory of gravitation. On the enormous scale of the phenomena we are considering, gravitation is the only force which we need to consider. But in addition to our theory of gravitation, which would of course admit an enormous number of possible motions of the galaxies, we need some observational facts to decide which out of the many cosmological theories devised by the mathematicians is the one describing the real universe. If we accept general relativity and if we accept the twin assumptions of isotropy and homogeneity, then we find that it is necessary for observation to give us only two numbers in order to completely specify the past and future behaviour of the universe. These are the Hubble constant (the distance corresponding to a given red-shift) and either the mean density or the deceleration parameter, which is a measure of the rate of change of the rate of expansion with time.

One can see how it is that one requires to know either the mean density or this deceleration parameter, because the only thing that is going to put a brake on the future expansion rate of the galaxies is the gravitational attraction of one for another. On a large scale this attraction will be determined by the mean density of matter, this being a measure of the amount of gravitating mass there is in the galaxies, and perhaps of any other matter that might exist in between galaxies. Indeed there is a very simple relation between the mean density and the deceleration parameter indicating that the link between the density and the deceleration is through the gravitational interaction.

We must introduce a further complication at this point. Most people know, in a vague sort of way, that the general theory of relativity is a theory in which gravitational interactions are due to a deformation of the geometry of space-time around gravitating bodies, and that the motions of neighbouring bodies are affected through this deformation of the space-time continuum in which they are embedded. One would therefore expect that the effects we have been talking about would be reflected in the geometry of the universe. In fact one can show that if the mean density of the universe is below a certain *magic density*, then the gravitational self-attraction of matter is not sufficient to put a brake on the expansion; in this case the geometry is said to be 'open' or hyperbolic. If the density is greater than the magic density, the expansion will slow down and reverse some time in the future and the geometry in this situation is 'closed' or elliptical. If by a lucky chance the density of matter is exactly equal to the magic density then the geometry appropriate to the universe as a whole is Euclidean and one is relieved of the task of learning anything of the complexities of non-Euclidean geometries!

We may look then on the two parameters we have to determine observationally, Hubble's constant and the mean density or deceleration parameter, in the following way. Hubble's constant is the scale factor which gives the scale of distances in the universe, that is, the distance corresponding to a particular red-shift between objects, while the mean density or deceleration parameter select that particular cosmological model which corresponds to the real universe. Hubble's constant gives the *order of size*, while the deceleration parameter gives the *type* of universe we live in.

Soon after his discovery of the relationship between velocity and distance for the galaxies, Hubble and others developed the theory of the observational tests among the possible models, and showed that there were three promising ways of determining the deceleration parameter. I propose to devote the rest of this chapter to a discussion of these

three methods, and to indicate what sort of progress we
have made with them, and what sort of progress we can
hope to make in the near future by exploiting the best techniques available today. I shall not say anything further on the
determination of Hubble's constant, except to remark that
twenty years ago it was considered to have the value of
550 km/sec/megaparsec (where the megaparsec is the
distance unit of extragalactic astronomy, and is the distance
travelled by light in about three million years). The favourite
value at present is 50 km/sec/megaparsec, which, if nothing
else, reflects the rapid and radical change in our understanding of variable stars and other distance indicators.

The three observational tests devised by Hubble and his
collaborators are, in order of importance: the red-shift—
apparent luminosity test; the red-shift—angular diameter
test; and the number—apparent luminosity test.

I have already mentioned Hubble's discovery of the
relationship between velocity (or red-shift) and distance, as
measured by apparent luminosity. To fix our ideas, let me
refer to Fig. 2 which shows a plot of the red-shifts of a large

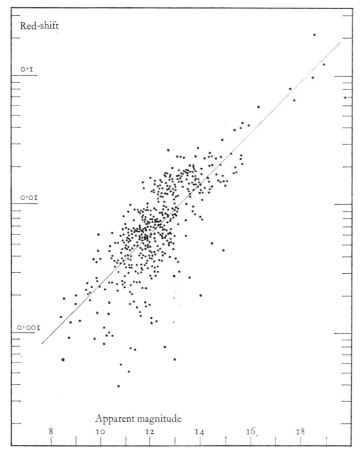

*Fig. 2: A plot of
red-shift against
magnitude for a
number of different
galaxies.*

number of galaxies against their apparent magnitudes, apparent magnitudes being simply related to the logarithm of the apparent luminosity. A mixture of different types of galaxies, both ellipticals and spirals, are plotted on this graph, and Hubble's law is represented by the straight line, giving a strict proportionality between velocity and distance. As you will see, it requires a considerable effort of imagination to connect the observational points with the straight line! But this lack of agreement should not surprise us if we recall that the intrinsic luminosities of galaxies of different sizes and types vary over wide limits. The sample of objects plotted as points on the graph is much too diverse to show the effect in which cosmologists are interested. We have somehow to find a way of restricting ourselves to those galaxies which are suited to our purpose both in having closely similar intrinsic luminosities and in enabling us to sample up to as great a distance as possible. This implies that we must select those galaxies with the greatest intrinsic brightness.

We are helped in the search for these bright galaxies by an interesting characteristic of galaxian distribution. It has been found that the vast majority of galaxies are not isolated in

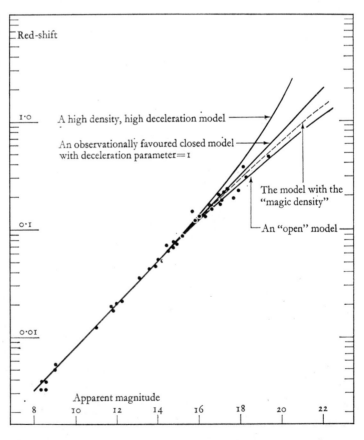

Fig. 3: A plot of red-shift against apparent magnitude for the brightest members of large clusters of galaxies.

Red-shift

1·0   A high density, high deceleration model

An observationally favoured closed model
with deceleration parameter = 1

The model with the "magic density"

0·1

An "open" model

0·01

Apparent magnitude

8   10   12   14   16   18   20   22

space but exist in clusters whose galaxy population varies
from about ten or twenty, such as the group in which our
own galaxy is embedded, to aggregates of many thousands
of galaxies. It has been found that if one selects the very
brightest member of one of these latter type of clusters, an
example of which is shown in plate 1, then one obtains a
red-shift—apparent luminosity diagram as in Fig. 3 which
is much more linear. It is immediately obvious that these
objects must be of closely similar intrinsic luminosities and
they define the linear Hubble relation between red-shift and
distance beautifully.

How can we use these observations to select one out of the
many possible cosmological models as the one our universe
most closely resembles? Well, since the velocity of light is
finite the photons our telescopes receive from distant
galaxies as we measure their apparent brightnesses are due
to light which left those galaxies at some time in the past.
Indeed even the sunlight that we receive on the earth's
surface records the state of the sun not at the present instant
but as it was eight minutes ago. The galaxies we are observing are at such distances that we measure the time that their
light takes to travel to us not in minutes but in millions of
years. The most distant galaxy for which a red-shift has been
measured is called 3C295, and for this object the time that
light from it takes to travel to us is getting on for 10,000
million (10^{10}) years. If galaxy 3C295 had ceased to exist just
before the earth was formed, we would not be aware of the
fact for some time to come! When our telescope and
spectrograph measures the red-shift and therefore the
velocity of such an object, it is measuring its velocity not at
the present time but at some time in the distant past. So if
galaxies were receding one from the other at a faster or
slower rate at that time, the velocity of the object as we now
measure it will be higher or lower than we would expect
on the basis of the relationship between velocity and distance
we measure for nearby objects. If the expansion rate was
(say) slower in the past, this would imply that Hubble's
constant was smaller in the past. The plot of velocity against
distance would then not be linear but would be curved.
On top of the observational points of Fig. 3 I have plotted
curves predicted by a number of cosmological models;
notice that the high density/high deceleration parameter
models show high curvature, indicating that in these models
the expansion velocity must have been very much higher in
the past and the expansion is braking hard. I have also
shown the curve for a model in which the expansion rate
was essentially constant in the past, this being a model of

such low density that the gravitational self-attraction of the matter of the universe is only sufficient to brake the expansion rate by a small amount. In all of these models of the universe there must have been some state in the distant past in which the matter at present condensed into galaxies was collected together in a very small volume. The matter must have exploded outwards from this state, starting off the expansion that we can see continuing today.

Unfortunately, when you look at Fig. 3 to discover which of the possible models predicts the correct shape for this plot of red-shift against apparent magnitude, you find that the observations do not really suffice to distinguish clearly between any but the most extreme models. It is obvious that if we are to make further progress towards an accurate determination of the deceleration parameter, we need many more observations so that averaging can reduce the errors, and more important still, we need observations of the red-shifts and brightnesses of galaxies at greater and greater distances. Fig. 3 shows that as the distance increases so the predictions of the different models diverge, and it becomes easier to make an observational choice from among them.

There are of course numerous technical difficulties involved in assembling the material for such a test, and in interpreting it correctly. I will discuss a few of these to give some idea of the uncertainties. From the first two photographs of galaxies shown on plate 1 it is clear that there is a problem in defining exactly where the image of a galaxy ends. The galaxy consists of a bright central nucleus surrounded by material whose brightness drops off as one moves away from the centre. The edge of the galaxy as we see it on a photograph does not really correspond to where the stars in the galaxy end, but it is fixed by the relative brightness of the outer parts of the galaxy and the night sky. The light of the night sky masks the faint outer regions of galaxies and prevents one from measuring the light output from the galaxy as a whole. The apparent luminosity must be corrected for this effect.

Another problem of a more theoretical nature concerns the question of whether galaxies evolve in brightness during those parts of their lifetime at which we observe them. One would naturally expect the brightness of a galaxy to change with time, because we know that it is a collection of stars, and we also know that stars change their brightness over long periods of time. The problem with calculating the evolution of galaxies is that one must add up the evolutionary effects due to its component stars, some of which will have one evolutionary history and some of which will have another. Calculations of these kinds of effects have not proved very successful; indeed we are not quite sure as to whether a galaxy gets brighter or dimmer as it gets older. Why should this matter from the point of our cosmological test? Because we have assumed that our standard test objects, the brightest galaxies in rich clusters of galaxies, are not only of the same intrinsic brightness today, but have been of the same brightness over a time that light takes to travel from the most distant of them, that is about 10,000 million (10^{10}) years. For if their brightness changes over this period, then as we observe the most distant objects at an earlier time in their history than nearby objects (whose light has only recently left them) we would not be really using a set of galaxies of the same intrinsic brightness. There would be a spurious curvature in the graph introduced by evolution which would confuse the choice we have to make between the possible models of the universe. To give some idea as to the uncertainty in the observational determination of the deceleration parameter due to our ignorance of how galaxies evolve, it is estimated that at present we would be hard put to it to distinguish between the cases of 'open' or ever-expanding universe and the 'closed' or ultimately contracting universe, even if there were no other source of uncertainty in the observations or their interpretation. This is disappointing, but our knowledge of stellar evolution and of the proportions of various types of star making up the galaxies is advancing rapidly, and one can see the day when this barrier to an interpretation of the cosmological test will be removed. For those who like to have some kind of answer to every problem, let me say that all available evidence now points to the deceleration parameter being somewhere between zero and two. The magic density that divides the two major classes of 'open' and 'closed' models for the universe is equivalent to deceleration parameter = a half, so you can see that the present results are tantalisingly unable to give us a decision between the two.

So much for the present status of this first of the three observational methods of choosing among the models of the universe. I will move on now to the second, that is, the angular diameter—red-shift test, which has not received as much attention as the first test in the past but which is attracting more interest due to the evolutionary uncertainties which seem to have temporarily stymied interpretation of the red-shift—apparent magnitude test.

One can obtain a simple idea of how this second test works by thinking of the following analogy. Consider a long, straight road stretching away into the distance with a man walking briskly along the road away from us. As he goes

C

away let us measure his apparent angular size from head to toe as he reaches regular distance intervals along the road. If we now plot the angle he subtends at our eye against his distance from us, we find that the angular size is inversely proportional to his distance from us. This is the Euclidean model. I have already mentioned that the mean density of matter in the universe was reflected in the deceleration of the universal expansion rate, and that it equivalently determined the type of geometry that distances between galaxies obeyed. If I now, instead of measuring our man on the road, measure the apparent angular size of galaxies at different red-shifts, I can test the nature of the geometry of the universe by seeing the rate at which they diminish in size with distance. One might expect angular size to be roughly inversely proportional to red-shift. If the size goes down in any other way, then the geometry of space must be non-Euclidean.

One might wonder what is the advantage of proceeding in this way by measuring sizes rather than apparent brightnesses. In fact there is little advantage in doing so if one confines oneself to the observation of the sizes of single galaxies, because our ignorance of how the total light output of a galaxy varies with time is paralleled by our ignorance of how its apparent size varies with time. For as a galaxy dims, its apparent size drops, and as it brightens its apparent size increases. Clearly the test is going to be difficult to interpret if it is a plot of the sizes of a number of galaxies all seen at different times in their evolutionary histories and therefore, perhaps, of different intrinsic sizes.

There is however a class of object which it may prove possible to use in our test, and against which these objections could not be advanced. This consists of whole clusters of galaxies considered as single objects. Now these objects, composed typically of many hundreds of galaxies, appear to be scattered uniformly through the universe, and to show a considerable similarity in properties. If one confines one's attention to those clusters of galaxies which show rough spherical symmetry, then it is found that, for a given population, the clusters seem to have roughly similar sizes. So instead of plotting a diagram of the angular diameters of single galaxies against their red-shift, one can plot a diagram of the angular diameters of clusters of galaxies against their red-shifts. Now whereas in the former case, the interpretation of the diagram in terms of the cosmological models was uncertain because of possible evolutionary changes, in the latter case it is hardly likely that the dimensions of a whole cluster of galaxies could have changed significantly during the light travel time between them and

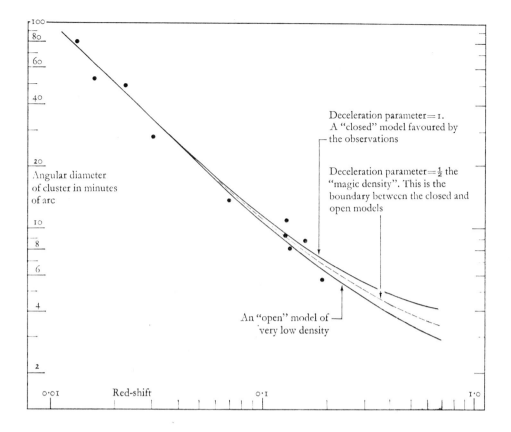

Angular diameter
of cluster in minutes
of arc

Deceleration parameter = 1.
A "closed" model favoured by
the observations

Deceleration parameter = ½ the
"magic density". This is the
boundary between the closed and
open models

An "open" model of
very low density

Red-shift

us. One of these clusters of galaxies, the so-called Coma
cluster, is shown in plate 1. This cluster is largely composed
of elliptical galaxies, which are the fuzzy patches of light—
easily distinguishable from the stars, which come out in a
photograph as sharp points. The cluster itself stretches
beyond the limits shown in this photograph, and has a total
diameter of about three megaparsecs. If one counts the
number of galaxies per unit area over the surface of the
cluster on a photograph taken with a large telescope, one
finds that there is a characteristic density pattern for each
cluster, with a high density of galaxies at the centre falling
off at the edges. One can define an angular size for the
cluster from this characteristic pattern, and plot this against
the red-shift of the cluster. A typical diagram is shown in
Fig. 4. As you will see, the observational points show somewhat more scatter than those of the previous test we used in
the preceding graph, but against this disadvantage we can
set out a high degree of confidence that the clusters we are
using have not changed appreciably in size during the light
travel time. The diagram also shows the prediction of a few
possible cosmological models, and it is obvious that the data
is not yet sufficiently extensive nor does it stretch to

*Fig. 4: Red-shift
plotted against the
angular diameter of a
number of clusters of
galaxies.*

sufficiently high red-shifts and distances to enable us to choose unambiguously between the models.

There is one further test that an optical astronomer can apply among the models. This is the third test on our list, that is, the observation of the way that numbers of galaxies mount up as one observes to successively larger and larger distances. The principles of this method are identical with the method of radio source counts used by the radio astronomers and described elsewhere, except that here one would compare counts of galaxies observed in visible light at various limits of apparent brightness. Apart from an attempt by Hubble to use this technique before the last war, it has received hardly any attention at all. This is largely because of the technical difficulties of measuring the apparent brightnesses of galaxies in large numbers.

The approach to observational cosmology we have been discussing here has a number of drawbacks. We must restrict ourselves to the use of astronomical objects that we understand fairly well, and that have a very high degree of uniformity. This is because small changes in (say) their brightness or their apparent sizes with velocity due to possible cosmological effects must not be masked by larger variations in brightness and size among the objects themselves. This restricts us to the use of certain kinds of bright galaxies, the giant elliptical galaxies that we think we understand fairly well. It also restricts us to fairly nearby regions of space because our existing telescopes cannot see these objects at greater distances. As we have pointed out, it is at large distances that the predictions of the various models diverge most strongly. It would be wonderful if one could use quasars in this way, because their red-shifts are very much larger than that of any galaxy yet observed. But it is not yet certain that these large red-shifts are to be interpreted as due to a velocity of recession, an interpretation we are confident of in the case of the galaxies. And even if it is one day demonstrated that the quasars are indeed at very large distances, their intrinsic brightnesses would vary so widely as to mask the effects we are looking for.

These drawbacks are offset to some degree by an advantage that optical astronomers have over their colleagues in other branches of the subject. The only practical way to select a precise model of the universe is by using our three tests. Other approaches can exclude some models and set limits on possible universes. But the approach through the three tests applied to optically observable galaxies can potentially give us a precise answer to the question as to what sort of universe we live in.

LOOKING
WITH NEW
EYES
Professor
Sir Martin
Ryle
FRS

Light from the distant galaxies cannot tell cosmologists everything; for one thing it is all too easily obscured by the dust clouds that interlace the spiral arms of our own galaxy. Fortunately, every hot object emits not only light, but radio waves. For example, the sun is a weak radio transmitter. Supernovae are powerful radio transmitters, but certain galaxies broadcast with such incredible intensity that they can be 'seen' with the new eyes of radio telescopes across distances so great that their light is swallowed up in the vast reaches of space involved. Mere heat could never produce these transmissions. What does? Now the BBC generates radio waves, but not because it is hot! Radio galaxies use a similar mechanism: the acceleration of electrons in magnetic fields. Accelerations produced by internal cataclysms, for where radio galaxies can be seen, they often appear to be in distress.

Professor Sir Martin Ryle, director of the Mullard Radio Astronomy Observatory, Cambridge, is Astronomer Royal. L. H. J.

We have all noticed that in a thunderstorm we hear the thunder some seconds after we see the lightning flash that causes it. This, of course, is because sound travels more slowly than light—at about one-fifth of a mile per second, so that if there is a delay of say ten seconds between the flash and the thunder, we know that it was about two miles away. Well, I suppose every schoolboy has done this kind of calculation, but not many of us bother to go on and interpret the fact that even the *light* doesn't reach us instantaneously. It travels about a million times faster than sound, so when we see the flash we are actually seeing something which has already happened about ten millionths of a second earlier.

Obviously, with things as close to us as a lightning flash, this time-delay between what we observe and what is actually happening doesn't make much difference to us. We can safely ignore it, and most of us do. But if we were timing the occurrence of one of the rare, sudden eruptions on the sun—the solar flares—we would have to remember we were seeing something which happened about eight minutes earlier. If we observed a supernova explosion half-way across the Milky Way system we would be seeing the death of a star about 25,000 years ago. And if a similar explosion happened today in the Andromeda Nebula, we could know nothing about it for another two million years.

The fact that it is impossible to get an up-to-date picture of distant objects may seem a disadvantage in astronomy, but in practice we make use of this delay to provide the most important key to cosmology—the study of the history of the universe on the largest scale. Our eyes and our

Chapter 3

instruments tell us that the sky is populated with stars, galaxies and clusters of galaxies as well as radio sources of various kinds. Where did they come from and how are they evolving? What we should like to do is to take a photograph, for example, of the Andromeda nebula and compare it with ones taken say 2,000, 4,000 and 6,000 million years ago, so that we could see how the nebula and the stars in it have changed. This, of course, we cannot do but we can do something nearly as good by comparing nearby galaxies with more distant ones. The light which is now reaching us from a distant galaxy is telling us how that galaxy looked when the light was emitted—hundreds or thousands of millions of years ago. In this way we might also hope to be able to answer that most fundamental question—whether the apparent expansion of the universe means that everything we now observe had its origin in a dense hot universe only 10,000 million years ago, or whether we are living in a universe which will go on without change, for ever—where the space left by the receding galaxies is replenished by the birth of new matter, which forms new galaxies.

For all these problems we need to be able to look as far away—and so as far back into the past—as we are able. This is why astronomers are always seeking to build even larger telescopes. Not to see further for its own sake, but to see further back in time.

Up until 1950 all that one could do was to study the light from distant galaxies with optical telescopes and try and sort out what kind of galaxies they were and how they were distributed in space. None of this, unfortunately, told us very much about how galaxies were formed nor did it allow us to decide whether the universe was changing or not. The difficulties were, first, that galaxies are very complex and the stars which provide nearly all the light only come into existence long after the galaxy is formed. Secondly, the optical observations just didn't reach far enough out into space—or back in time. Thirdly, galaxies tend to occur in clusters. This doesn't worry us too much in studying distant ones, but makes it difficult to get a proper sample of nearby ones for comparison. It's just that when we look out over very large cosmological distances, we see a large number of clusters of galaxies. So if you want to estimate the average value of the density of galaxies in space, you have a large sample to work with. You obviously cannot do this when you are making relatively local observations, over a volume of space comparable with that of a cluster of galaxies. Here the average density would entirely depend upon whether our galaxy were inside a cluster or outside.

But in 1950, something happened that enabled us to cut through some of these difficulties and restrictions. This was when one of the first remarkable results emerged from the new science of radio astronomy. The radiation emitted by the universe is not confined to those wavelengths which our eyes can detect—it would be a remarkable coincidence if it were. Cosmic radiation extends both to shorter wavelengths such as ultra-violet and X-rays, and to the longer infra-red and radio waves. In 1950 one of the most intense sources of radio waves, discovered a few years earlier, was identified with a faint galaxy no less than *500 million light years* away. This immediately suggested that radio astronomy might be able to detect galaxies much further away than optical astronomy. Not only that, but later observations showed that these 'radio galaxies' did *not* form those awkward clusters. So it began to look as if radio astronomy might prove to be an extremely useful tool for finding out whether the universe was evolving or whether on a large scale it was unchanging.

During the next decade progress was rapid, both in the detection of large numbers of sources and in the association of some with faint galaxies. During the 1960s *some* radio sources were identified with quasars—those intensely bright star-like objects whose very large red-shifts seem to imply that they are at great distances away. Like that associated with galaxies, the radio emission from quasars usually comes from two components—one each side of the optical object, which suggests that the emission may well be due to clouds of plasma being shot out from some great explosion. In fact, from radio evidence alone, it is difficult to see any clear-cut distinction between radio galaxies and quasars, and this suggested that a quasar might well be the nucleus of a galaxy, which we just happen to have caught at an early stage of development—a stage in which the activity responsible for the radio components is intense. More recent observations show that many sources, identified with both galaxies and quasars, also have a central *radio* nucleus, confirming activity in the nucleus does indeed occur in both classes of source.

Now, there's a big argument still going on about whether the quasars *are* in fact at the great distances implied by their red-shifts—they are certainly remarkable objects and the shift of the spectral lines *might* be attributable to some other mechanism. For example, if a quasar contained some very massive, highly condensed object, this might produce a gravitational red-shift. After all, if you look at the spectrum of sunlight you find a very small shift of the spectral lines towards the red—which is due to the effect of the sun's gravitational field on the light leaving its surface. With a more massive highly condensed body this gravitational red-shift would be much greater. This is a possibility, but so far nobody has proposed a model that is capable of explaining the details of the observations.

Fortunately these arguments do not affect our ability to use radio source studies in cosmology—because there is an entirely independent proof that most radio sources are powerful sources at great distances. And this proof does not depend on optical evidence at all. In fact, even if some of the quasars were relatively local with a different mechanism for the observed red-shift, then they could only represent a small fraction of the sources observed and would not affect the cosmological conclusions.

The proof is based on very simple observations; first the number of sources which we find with different intensities, and second the radio brightness of the sky. The arguments are the same as those put forward as long ago as 1839 by the German astronomer Olbers (described in chapter 1 by Sir Hermann Bondi).

The dark night sky remained a paradox until it was finally solved in the early part of the present century by the discovery of the red-shift—that the universe is in a state of expansion in which the greater the distance of a galaxy, the larger its red-shift. What happens is that as a galaxy recedes further and further from us, the light becomes both redder and weaker. So the most distant galaxies contribute hardly anything to the brightness of the night sky. We look in their direction and it is dark.

This same reasoning can be applied in a slightly different way, to the observations of the radio sky background. As you might expect we find there is enhanced radiation from the plane of our own Milky Way. But we can also measure the radiation from extragalactic space. There may be other types of source contributing to this background besides the radio galaxies and the quasars—but we can certainly establish an upper limit to what they contribute.

Now, with our radio telescopes, we find, say, 1,000 sources brighter than a given intensity, but we don't know their distances. If we suppose them to be of relatively low power and situated at small distances—as some people have proposed for the quasars—then there must be many more distant ones which, though we can no longer detect them individually, must contribute to the background radiation from the sky. But, of course, we know the strength of the background radiation so we can set a lower limit to the average distances of our 1,000 sources. In practice, with the deepest source surveys we find that the radiation from all the individual sources we can detect, represents a major part of the total sky brightness—so that we are detecting sources over an appreciable fraction of the 'observable' universe.

This means we know that some of the sources we have already observed are at distances approaching the fundamental limits of observation, so the study of their distribution in space—and any changing properties which they may have at different distances—can be used to explore the history of the universe on the largest scale.

Well, what happens when we do this? The first, and most remarkable result of all, concerns the numbers we find of radio sources of different intensities. As we proceed outwards from the most intense—and presumably nearest sources, we find a great excess of fainter ones. Now this suggests that in the past either the power, or the space density of the sources was greater than it is now. Whichever way it is, the universe must have changed radically within the time-span accessible to our radio telescopes.

But at still smaller intensities we find a sudden reversal of this trend—a dramatic reduction in the number of the faintest sources. This convergence is so abrupt that we must suppose that before a certain epoch in the past, there were no radio sources. Both these observations, therefore, seem to indicate that we are living in an evolving universe—which has not always looked the same.

If we try to estimate the time-scale of the evolution of the radio sources from the physical properties of the nearby ones which we are beginning to know something about, it seems that we are looking back through 9/10ths of the age of the universe. The abrupt reduction in the number of very faint sources corresponds to an age about 1/10th of the present age. This is probably associated with the actual formation of galaxies from the primeval gas because —certainly according to our present knowledge—there seems no way of producing radio sources before we have galaxies.

So the picture presented by the radio source observations supports the idea of an expanding universe which evolves with time, from an initial state of very high temperature and high density. One of the great discoveries of recent years has given evidence of this extreme phase. This evidence also comes from radio observations—a weak background radiation which predominates at centimetric wavelengths and is the type one would expect to find if it were the emission during the first few hours of the present phase of expansion.

As the expansion from this high temperature, high density

state proceeds the gas cools, and after some 1,000 million years it begins to condense into galaxies. Subsequently, with the formation of stars, nuclear sources of energy become available, and eventually the gravitational energy which we believe to account for the pulsars. A similar source of energy may be responsible for the great explosions which give birth to the radio galaxies and quasars.

It's a great stroke of good fortune that there are these intensely powerful sources of radio waves distributed throughout the universe. It is only these that have made it possible for us to undertake this initial exploration of a very large part of the time-span of the universe.

What we now need is an understanding of the physical mechanisms involved in the formation of a galaxy from the primeval gas, and its subsequent evolution from this earliest stage to that involving the sudden enormous energy production apparent in radio galaxies and quasars. The discovery of these mechanisms is likely to depend mainly on better observations at radio wavelengths, but will also require further optical work as well as observations from spacecraft at infra-red and X-ray wavelengths.

The new 5 km radio telescope built at Cambridge was designed specifically to improve the resolution, or mapping detail available at radio wavelengths, and is the first instrument to provide radio maps showing the same sort of detail as that available in the best optical photographs. With these maps it is possible that we shall be able to understand the complicated processes involved in the development of a radio galaxy, and explore the way in which these have changed during the history of the universe.

SOURCES
OF COSMIC
POWER
Professor
Donald
Lynden-Bell

Candles give light because of the energy emitted when tallow combines with oxygen; electric lights glow because electricity is forced through a fine wire; but what keeps the stars and the galaxies shining through the empty night of the universe? Whatever the source of power, it must be unimaginably immense. Yet quasars, those star-like objects that may lie near the edge of the universe, can outshine entire galaxies. Their energy output is of the order of 10^{61} ergs or ten million million million million million million H bombs! Heat generated by the galaxies accounts for as much energy as light. Finally take into account the X-rays, gamma rays, radio waves etc emitted and the total energy budget of the universe is found to be stupendous.

What is the source of all this energy? Donald Lynden-Bell, Professor of Astrophysics at the University of Cambridge, illustrates the pitfalls that can beset the unwary theoretician in his search for an answer. L. H. J.

In order to hold up its outside, a star like the sun must have central regions which produce a pressure sufficient to balance the great weight of the overlying layers. How does the sun do it? Matter can withstand high pressures either by being very dense or very hot, and since the average density of the sun is only the same as that of water, one deduces that its centre must indeed be hot. Following these ideas through, Eddington found that the centre of the sun has to be at between ten million and forty million degrees centigrade. The question then arose: how does the sun sustain such a high temperature? On consulting his colleagues in physics, Eddington was informed that, high though the temperatures might be, they were too meagre to provide reactions between the nuclei of atoms, so nuclear power was out. Other fuels were already known to be inadequate, so, doggedly ignoring his colleagues' advice, Eddington still considered that the most likely source of the sun's energy was some chain of nuclear reactions—one, in fact, in which the nuclei of four hydrogen atoms were fused together to make one helium nucleus. Eddington's reason for choosing this reaction was that hydrogen has the simplest nucleus, the proton, and both hydrogen and helium are common in the sun and stars. Indeed, the atomic masses of all the elements are approximate multiples of the mass of hydrogen; an observation that has far-reaching implications when one comes to ask where the atoms of these elements came from. I say approximate because, for example, the mass of four hydrogen atoms is a little more than the mass of one helium atom but the small difference of only three quarters of a per cent nevertheless can be changed into a considerable

Chapter 4

amount of energy. The exact amount is given by Einstein's relation $E = mc^2$, where E is the energy in the mass and c the velocity of light. Even in the 1920s, before the advent of the hydrogen bomb, this nuclear fusion process seemed a likely power source for the sun and stars, but at that time the details of the process were quite unknown. When Eddington was asked, he merely pointed out that the elements seemed to be built up out of hydrogen and if the centres of stars were not hot enough for this nuclear fusion, could he be shown a hotter place?

The close connection between the making of helium and the origin of sunshine in the energy so liberated, suggested that all the other shining objects in the universe might be powered by nuclear energy and that, by further fusion reactions, all the elements might be made in the process. After all, stars like the sun do not contain hydrogen and helium alone, and the relative abundances of the elements in stars can be determined from the details of their spectra. By analyses of these, astronomers find that the outsides of the sun, most stars and the gas clouds between them, are made of a standard mixture. This is about 66 per cent hydrogen by mass, with about 30 per cent helium and a per cent or two, no more, of all the heavier elements combined. In a triumphant piece of astrophysics, Geoff and Margaret Burbidge, William Fowler and Fred Hoyle demonstrated that nuclear reactions that have been studied in nuclear reactors on earth would, if assumed to take place in stars, predict many of the details of these observed abundances of the elements heavier than helium.

When did the element building take place? The stars of the globular clusters and the associated stars that are passing with high speeds through the neighbourhood of the sun are known to be old. Many would argue that they are the oldest in the galaxy and their outsides have been left untouched for untold aeons. On their surfaces we see, as though fossilised, the sort of gas out of which these stars must have condensed, and, sure enough, the higher elements are missing or almost missing. However, there are other stars that are either their contemporaries or almost as old, in which we find the normal complement of the higher elements, such as carbon, nitrogen, oxygen, etc. We deduce that these higher elements (of which we ourselves are mainly composed) were built in the early youth of the galaxy, or even as it formed. Whether we can say the same of the much commoner element, helium, is another matter. Unfortunately, helium is difficult to detect because it needs a high temperature to excite its spectrum. The old stars are too cool to make helium radiate its spectral lines at all, so we cannot tell directly whether those stars in which the higher elements are almost missing, have their 30 per cent complement of helium, or whether it is missing too. Two indirect arguments suggest that helium is present even when the higher elements are not. Firstly certain stars in the old globular clusters, whose outsides are very deficient in the higher elements, vary their light as they pulsate in and out. The mechanism that drives the pulsation is now understood and involves a substantial amount of helium not far below the surface of these stars. Secondly two galaxies have been found full of hot young stars, which show a deficiency in the higher elements, but have the normal complement of helium. Both these observations imply that the bulk of helium may have been formed before the galaxies existed. We therefore need a theory for the origin of helium that can account for a substantial amount of its manufacture even before the stars were formed.

Eddington's challenge to find a hotter place than the centres of the stars was taken up in the 1940s by George Gamow in the United States. In his early studies of big bang models of the universe, he hoped that he could explain the abundances of the elements as a natural consequence of the high temperatures and densities during the initial bang. He found it quite possible to make helium, but that the material expanded too quickly to make substantial quantities of the higher elements. Although more modern physics has considerably revised Gamow's early calculations, his qualitative conclusions remain intact, and it now seems that a wide range of initial conditions during the big bang give rise to roughly the observed amount of helium. Gamow's theory of the origin of helium in the big bang has a very wide measure of support among astronomers today, but it is as well to remember that it does depend upon believing that astrophysicists know what was going on in the first fifteen minutes of the life of this universe. Here it is as well to remember the saying, 'Astrophysicists are often in error but *never* in doubt!'

Another approach is to say that some helium is certainly being made by hydrogen fusion in stars. The question is, how much? Bearing in mind that the higher elements too are made by fusion and that some energy comes from gravitational contraction, if we attribute *all* the light of all the stars to hydrogen fusion, then we shall slightly overestimate the rate of helium creation. If we then multiply this current rate of helium creation by the age of the galaxy, we get a slight over-estimate of how much helium may have been made in the stars over all galactic history. In this way we find that less than 4 per cent of the initial hydrogen would have been turned into helium by stars, while the observed total amount of helium is 30 per cent.

The currently accepted picture is that the bulk of the helium was made by Gamow's process in the big bang and that all the stars shining all the time have only added a little. In fact, the mass of extra helium they have made is probably comparable to the combined mass of all the higher elements. This standard picture has the material of the universe emerging from the big bang as a mixture of about 25 per cent helium and 75 per cent hydrogen. What happens then is that the galaxies condense out of this with no higher elements and finally the extra helium and all the higher elements are built in stars in the same processes that release the energy into starlight.

You may find my argument up to now quite convincing, but that does not necessarily mean it is correct! To show you what I mean, instead of tamely accepting these ideas, let us contemplate rejecting Gamow's big bang origin for helium and try to account for all helium in terms of starlight. In making the extra helium, much more energy would have been released, and to avoid apparent contradiction with estimates of the total rate at which the galaxy shines now, the extra energy released must have made it very much brighter in the past than it is now. This idea of a spectacular youth for the galaxy is not unnatural when you consider that the amount of energy to be lost when 30 per cent of the hydrogen in a galaxy is turned into helium is quite prodigious; enough to blow the galaxy to bits not once but ten thousand times. In fact, we should regard the hydrogen of which most of the universe is composed as a dangerous material. It has only to get very hot at a reasonable density for the fusion reactions to take place explosively as in the hydrogen bomb. Do we observe young galaxies that radiate at a hundred times the normal rate and which are racked by explosive outbursts? Quasars might well fit this description.

Quasars are more common out in deep space where we look back in time to days when the sun was still unborn and galaxy births were more common. To investigate further the idea that they are, as it were, vast hydrogen bombs, we need a greater knowledge of what quasars are like. We know that their spectral lines have large red-shifts which probably correspond to high speeds away from us due to the expansion of the universe, but we know neither their masses nor how long they live. At present, a promising approach is to find objects intermediate between quasars and galaxies whose

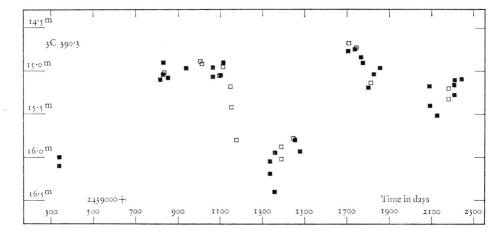

Fig. 5: Royal Greenwich Observatory recordings of the brightness of the variable N galaxy 3C 390.3, the brightness in magnitude B plotted against time in days.

basic properties are not so open to dispute. The discovery of quasars was preceded by the discovery of a remarkable class of bright galaxies, the N galaxies. In an N galaxy, *most* of the light comes from an apparently point-like nucleus, whereas a normal galactic nucleus is seen as a bright central knot which gives only a small fraction of the total light. N galaxies are found in clusters of normal galaxies that have the same red-shift (plate 4). These red-shifts are not particularly large, although some are amongst the largest red-shifts recorded for normal galaxies. Like quasars, N galaxies were originally discovered among the optical identifications of the third Cambridge catalogue of radio sources, but by now we know of examples of both types that are not strong in the radio region of the spectrum. Like quasars, N galaxies have spectra with more of the light in the ultra-violet and infra-red parts of the spectrum than normal stars give. This light cannot be stellar light. N galaxies and quasars vary their light output irregularly with time scales of the order of a year (as shown in Fig. 5), although in addition significant variations can be found over a week. None of them are as constant over a five-year period as the stars are.

When Fritz Zwicky of the California Institute of Technology first claimed to have discovered a galaxy whose light was variable, most astronomers did not take any notice. When quasars were first shown to be variable, it was still believed that these must be objects inside our galaxy because distant galaxies thousands of light years across could hardly change their light output significantly in one month. Now it has been realised that the light that changes is non-stellar light coming from very small regions, probably much less than a light month across, but the source of this irregularly variable light is unknown. Mystery still surrounds the ultimate source of the energy. It is not unnatural to suppose that quasars are distant and tremendously powerful examples of N galaxies in which the non-stellar light far outshines the surrounding galaxy. May we look upon such objects as the furnaces in which the bulk of the universe's helium might be forged? What do their wildly irregular light variations mean? They are too intense to be caused by individual supernovae of the kinds already known, but perhaps we should expect something rather special when a galaxy first turns from gas into stars. We *do* know that normal stars contract very slowly to light up their nuclear furnaces, but it was shown thirty years ago that this slow, stable lighting up would not be possible for stars of over a hundred solar masses. Here the fusion reaction would set off a violent instability. Professor McCrea, in a witty mixture of depth and simplicity, has asked, 'How does a cloud of gas know that it is too massive to form a star before it has tried?' His theory of quasars is based on the intriguing idea that massive clouds try to form stars but blow up like hydrogen bombs when they find themselves too massive to do so. Such explosions could well exceed the power of normal supernovae due to the greater masses involved. This picture of a quasar is an assembly of pieces but many astronomers consider that a single massive object is a more realistic picture.

What are the possible sources of the prodigious power of a quasar consisting of a single object? If we try to make a model based on hydrogen fusion, then the large total energy emitted, together with the fact that this can represent only three-quarters of a per cent of the total mass-energy, leads us to consider bodies of over a thousand million times the mass of the sun. Furthermore, if we take the observed rapid light variations as a measure of the time that it takes light to cross the emitting region, then this very massive body must be very compact, perhaps the size of the solar system. Now the gravitational binding of a large mass in a small region will be enormous. In fact, in this case, we find that the gravitational binding is stronger than the binding of the hydrogens into helium. This means that more energy was lost by radiation during the gravitational contraction of the whole body to this small size, than was lost in making the helium. Thus our model of a quasar based on hydrogen fusion as its major source of power has, instead, turned out to be a model in which most of the energy output has come from the gravitational energy released on contraction. Clearly, it would be better to start again with a new model based on this gravitational contraction as its major power source. When this is done, we find that we need less mass, since a greater fraction can be radiated away.

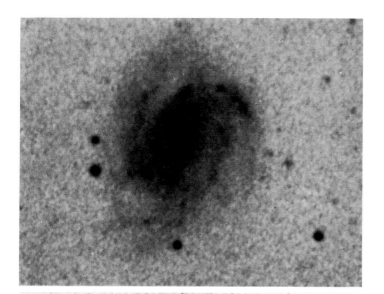

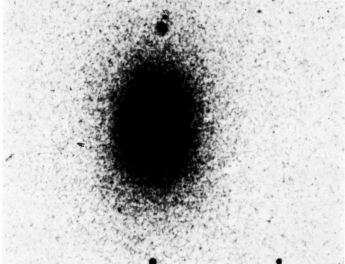

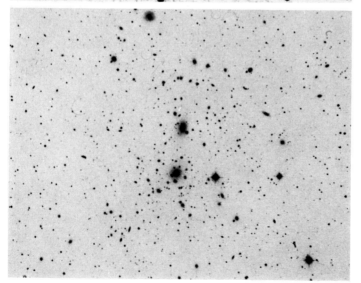

Top
Spiral galaxy
NGC 3780 is so far
away that the light
from it has taken 120
million years to
reach us. The image is
so faint that it
merges at the edges
with the brightness of
the night sky.

Centre
Elliptical galaxy
NGC 5322. The
photographic grain is
very pronounced,
obscuring the
position of the edge
of the galaxy, which
is almost certainly
bigger than this print
suggests.

Bottom
These galaxies in
the Coma cluster are
so far away that
individual stars are
not distinguishable.
The galaxies show
up against the
foreground of the
stars in our own
galaxy as fuzzy,
often elliptically
elongated, patches.

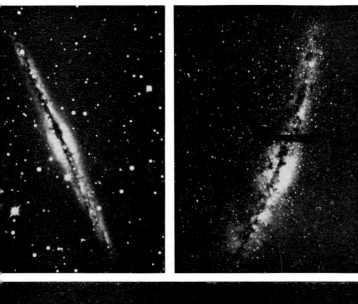

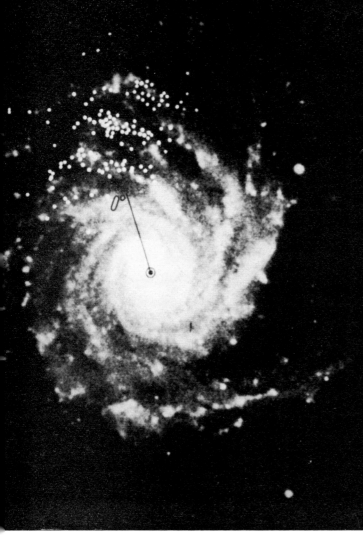

Far left
The Andromeda
nebula

Top left
This galaxy,
NGC 891,
has a remarkably
similar appearance
to our own, including
the dark band of
inter-stellar dust.

Top right
Photograph taken
with an all-sky
camera in infra-red
light to penetrate
some of the dust
clouds. It shows the
whole southern
Milky Way giving,
in effect, an edge-on
view of our galaxy.
Note the dark band
of inter-stellar dust.

Bottom
Galaxy NGC 1232
The white dots are
spiral features of our
galaxy as super-
imposed by Becker
(Basle University),
and give some idea
of how little we can
see of our own
galaxy because of
obscuring dust-
clouds.

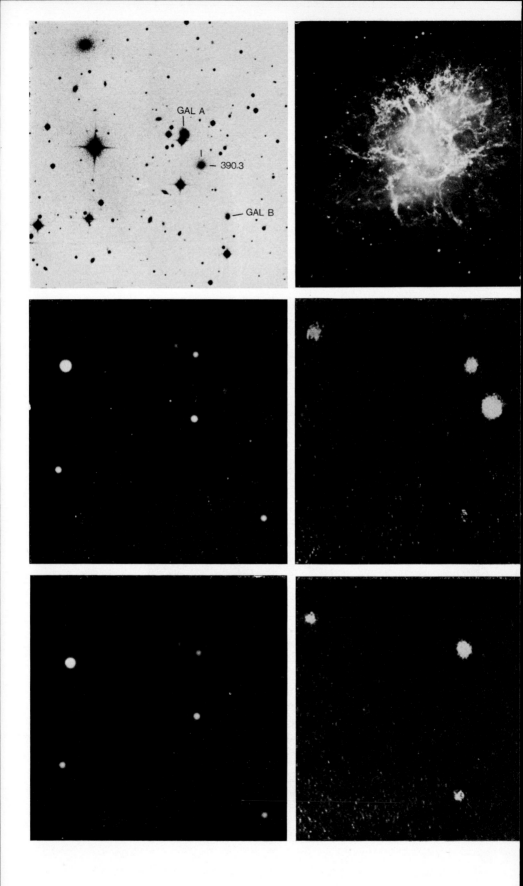

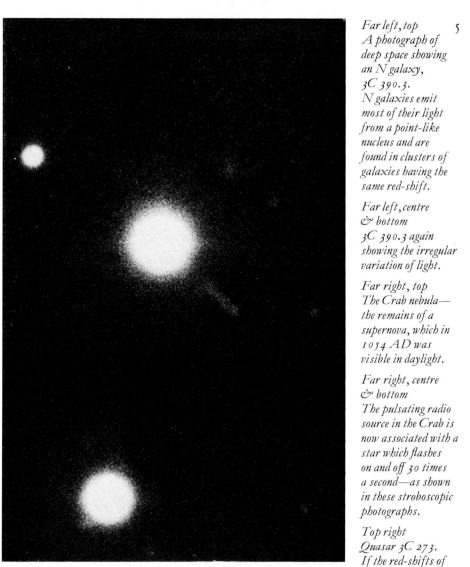

Far left, top
A photograph of
deep space showing
an N galaxy,
3C 390.3.
N galaxies emit
most of their light
from a point-like
nucleus and are
found in clusters of
galaxies having the
same red-shift.

Far left, centre
& bottom
3C 390.3 again
showing the irregular
variation of light.

Far right, top
The Crab nebula—
the remains of a
supernova, which in
1054 AD was
visible in daylight.

Far right, centre
& bottom
The pulsating radio
source in the Crab is
now associated with a
star which flashes
on and off 30 times
a second—as shown
in these stroboscopic
photographs.

Top right
Quasar 3C 273.
If the red-shifts of
these quasars are due
to expansion of the
universe, there are
probably too many
of them at large
distances to be
compatible with the
steady state model.

Bottom right
Starting with the
homogeneous
conditions of the
big bang it is
surprising that
the universe should
have evolved to this
highly inhomogeneous
clusters of galaxies.

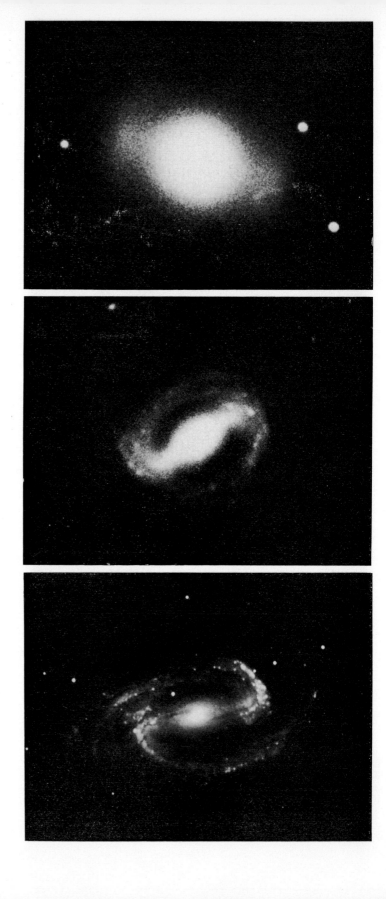

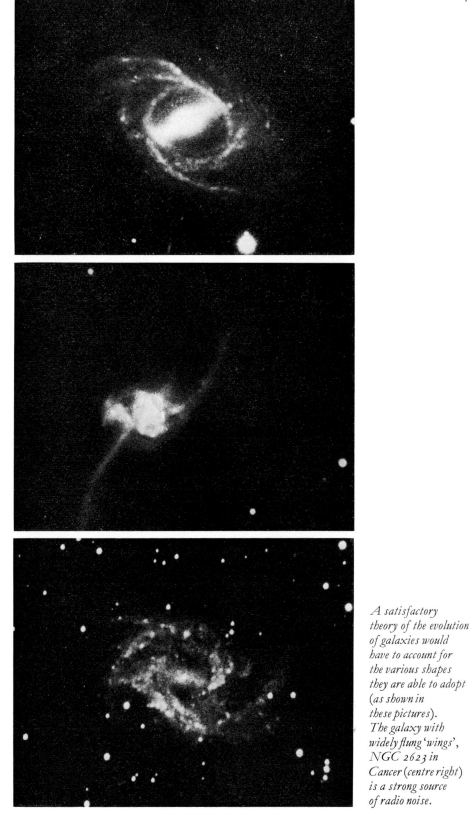

*A satisfactory
theory of the evolution
of galaxies would
have to account for
the various shapes
they are able to adopt
(as shown in
these pictures).
The galaxy with
widely flung 'wings',
NGC 2623 in
Cancer (centre right)
is a strong source
of radio noise.*

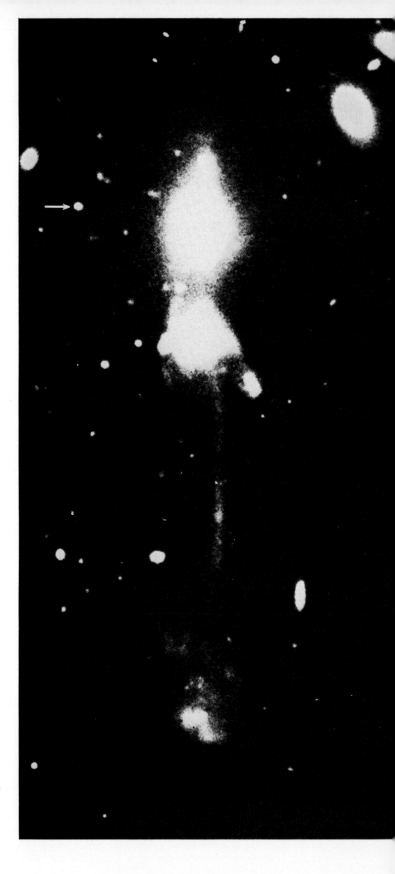

The arrow points to a quasar very close to a remarkably disturbed galaxy.

I have been particularly interested in the final fate of such bodies. It is clear that they have insufficient energy to re-expand and there does not seem to be anything that can stop them from steadily getting smaller and hotter. Many have argued that this process will finally lead to matter making a black hole, which pulls any surrounding matter inwards, as Professor Penrose describes in Chapter 9. Let us consider what happens as the body contracts. First, it will spin faster. Eventually, it will spin off a disc of material orbiting around the central mass. The central body will rotate faster than the disc it leaves behind and so any friction or dissipation in the system will lead to a transfer of the spin, or angular momentum from the body to the disc. In time, we may expect the central body to contract so far that even light has difficulty in escaping from its gravitational field. Eventually, light may not be able to escape at all and the central body will be unable to communicate with those outside. However, outside observers will still see the effects of a curved geometry like that surrounding any other massive object.

These ghosts of what was once matter have many of the attributes of death and hell. Going down into one is an uncomfortable one-way process, as Roger Penrose graphically describes, and any cries for help fall back to the afflicted rather than reaching the world without. The considerable mathematical difficulties surrounding the mathematical description of the making of black holes have been partially overcome by the work of Dr Bardeen at the University of Washington in Seattle, America. He demonstrates that it is in principle possible for a body to evolve into a spinning black hole. These as yet hypothetical objects are described by simpler mathematics so the properties that they would have can be fully studied. The spinning disc that was left behind in making one, may eventually lose its angular momentum to the rest of the galaxy and so work its way down towards the hungry black hole. One may calculate that the energy that becomes available to the outside world before any mass is finally swallowed can be up to 42 per cent of its rest mass. Thus, gravitational contraction can be 42 per cent efficient in converting rest mass into energy as compared with about 1 per cent for nuclear fusion. At such a high efficiency, only about one solar mass per year need be swallowed in order to power a quasar.

There is one way of getting a reasonable idea of quasar masses. This involves the notion that the blast of radiation coming from the quasar should not be far weaker than the gravitational attraction of the central mass. The observation that a few quasars have gas falling towards the light source,

49

whereas some have gas coming out towards us, leads one to speculate that the blast of radiation pressure is sometimes less, but sometimes greater than the gravity. If we set these quantities equal for a normal ionised gas of hydrogen, we get the Eddington limit for the energy generated per gram of central mass, equal to 60,000 ergs per gram per second. This is about 30,000 times the rate of energy production per gram in the sun. It gives the quasars masses of about a hundred million suns—very similar to the masses of ordinary galactic nuclei.

Our present picture, then, is that while stellar light is mainly powered by hydrogen fusion to make helium, nevertheless, most current theories suggest that quasar light is gravity powered from far larger masses. Whereas the first result is secure, the second is not strongly founded. If true, it still leaves us with only a fraction of the observed helium made by processes that we see now and we would conclude that Gamow's theory, that the bulk of it was made in the big bang, is an attractive one.

I had better make it clear that although they are a natural prediction of the general theory of relativity, black holes are at present a theoretical speculation. Although none of us yet feels that he knows for certain, some astronomers think that some of the objects observed have black holes in them.

We know already that some stars have collapsed to a size only ten times larger than that at which they would become black holes. Theory predicts that no such half-way-house is possible for the more massive stars. You will find a full description of what happens to these in Chapter 9, but the picture is so strange that I would like to repeat it here in slightly different form. If we consider the possible states of dead stars, we have firstly black dwarfs, that is, white dwarfs that have cooled off and become like massive stones. In a sense, this is just what they are, because if one piled more and more stones together and kept at it, the pile would grow bigger and bigger until it became planet-sized. It would continue growing until it reached about the mass of Jupiter, but on piling on still more stones, the extra mass would start crushing the atoms in the centre so hard that further additions would only make the body shrink. By the time the mass reached that of the sun, the pile of stones would be only Earth-sized—a white dwarf with a density of around a ton per cubic centimetre. Additions beyond the Chandrasekhar limit of about 1.2 solar masses would lead to a catastrophic change because inside such a body the electrons have little room between one and the next. The enormous pressure needed to prevent further collapse is

provided solely by the agitation necessary to obey a fundamental physical law known as Heisenberg's uncertainty
principle. However, when this process forces the electrons
to vibrate so fast that their energies become over three-
quarters-of-a-million electron volts, the electrons, rather
than being forced to take part in a still faster rat-race, go
into expensive retirement by combining with protons to
make neutrons. This process is the reverse of the everyday
one in which free neutrons decay into protons and electrons.
With further mass addition, the body is unable to withstand
its own gravity, so it falls together until it reaches the densities of atomic nuclei. Here it is the neutrons themselves
whose hard cores and Heisenberg agitation support the
star which is, by now, predominantly made of neutrons.

The prediction of such neutron stars was made independently by Landau in Russia and by Baade and Zwicky in
California, who also suggested the correct astrophysical
context in which neutron stars should be found. Zwicky
even suggested that there ought to be one in the Crab nebula
(plate 4), a prediction confirmed more than thirty years
later when, first by radio and then optically, the strange
star that Baade had always suspected, was shown to be
flashing thirty times in every second.

Before returning to the story of the discovery of neutron
stars, let us complete our picture of the stellar graveyards.
Briefly, theory predicts that above about two solar masses,
even the neutron stars cannot withstand gravity without
being hot and shining. The trouble is that extra energy is
involved in the agitation necessary to produce the pressure.
This extra energy itself has its equivalent mass and so the
weight is increased. At masses over about twice that of the
sun, the extra weight more than offsets the extra pressure and
so the body collapses. There seems to be no further stopping
place before the body becomes a black hole.

In 1956, I was still a student, but I remember hearing of
Dr Hewish's studies of the outer corona of the sun from the
fluctuations in the signals of radio sources such as the Crab
Nebula as they pass behind it. It was not long before he had
established that the sun's magnetic field was drawn out into
radial streamers and his was possibly the first detection of the
sparse solar wind that blows from the surface of the sun at a
speed of some 500 kilometres a second. It was Hewish's
persistence in the study of the twinkling of the radio stars,
and of the solar wind that causes the twinkling, that led to
the amazing chance discovery of the pulsars. It is now
agreed that these are the long-predicted neutron stars. As
knowledge of the unevennesses in the solar wind increased,

Hewish was able to use the amount of twinkling to measure the angular sizes of compact radio sources such as quasars. The principle is the same as that in visible light: stars twinkle but planets, with their larger angular sizes, do not. To record the twinkling and measure the sizes, Hewish had designed and built a large array of radio aerials, and because he wanted to get the details of the twinkle, he was not averaging the signals for the usual time. Late in 1967, a strange source of 'interference' occurred on the records. A research student, Miss Joscelyn Bell, found that this interference occurred again and again, not very strongly, and for just four minutes a day. It consisted of a series of 'pips', brief radio pulses separated from each other by one and a third seconds. The strength of the pips was very variable, but the timing was apparently perfect. As time went by, it was found that the pips did not occur at exactly the same time each day, but rather at the time that a certain patch of sky was overhead. This confirmed Miss Bell's belief that the 'interference' was not unimportant man-made noise, but came from the sky. It was not long before she had discovered another one and the race to uncover the mystery was on.

The accurate position in the sky was established by using the small variation in the arrival times caused by the earth's motion around the sun. A rough estimate of distance was determined from the amount by which the pips recorded at low frequency had been slowed by the interstellar gas, as compared with those at high frequency. At first there was some debate as to whether the one-second pips could not arise from some white-dwarf star that had been set vibrating, but such speculations were found wanting when the one thirtieth of a second pulsar in the Crab Nebula was discovered with the radio telescope at Jodrell Bank. Hewish had found the strange small source in the Crab some years earlier, but that was before any pulses had been discovered. So far this is the only pulsar known that also flashes on and off in the optical part of the spectrum. (This is beautifully demonstrated in the pictures taken stroboscopically at the Lick Observatory shown in plate 4.) The discovery that the star that Baade had always suspected was flashing at this fantastic rate, beautifully confirmed Zwicky's concept that neutron stars ought to result from supernova explosions, such as the one that produced the Crab Nebula.

It is now agreed that it is the magnetic field of the rotating neutron star that generates the pulses as it is turned over and over. Most theorists believe that the mechanism is related to the region some distance from the neutron star where the magnetic field would have to travel with the speed of light

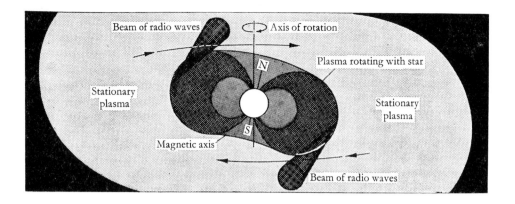

Beam of radio waves | Axis of rotation
Plasma rotating with star
N
Stationary plasma
Stationary plasma
S
Magnetic axis
Beam of radio waves

if it were to keep up with the rotation (Fig. 6). The rapidly changing magnetic fields in such a region generate large electric fields through Faraday's laws of electromagnetic induction. Any charged particle finding itself in such electric fields will be accelerated to velocities close to the speed of light. Thus, pulsars will be accelerators for cosmic rays; indeed, the shining of the Crab Nebula itself is powered by cosmic rays and the Crab pulsar is gradually slowing down at just the rate one would expect from the power required to keep the nebula shining.

Fig. 6: A possible mechanism for the action of pulsars.

Pulsars are about 10 kilometres in diameter, but take the same mass within 1 kilometre and the object would become a black hole. Can we find objects too massive to be pulsars and yet associated with very rapidly varying light, cosmic rays, etc? Three mysterious objects are particularly worthy of attention. First, there is the variable star and variable radio source B.L. Lacertae, which is probably extra galactic (it lies not very far from the galactic plane so it *could* be within the galaxy) but which has varied within 15 minutes! It has no lines in its spectrum, so there is no known red-shift. Then there are two X-ray sources, Scorpio XI and Cygnus XI, neither of which is understood, though both lie within the galaxy. We know most about Cygnus XI, because the invisible X-ray emitter swings about a normal bright star. Studies of the velocity of this star with the Isaac Newton Telescope show that the invisible X-ray source has an orbit corresponding to a mass of at least three times the mass of the sun. Is this mysterious object one in which the X-rays are being generated from material circulating about a black hole at near the speed of light? It has recently been discovered that the emission lines in the spectrum of Cygnus XI vary out of step with the others. They must come from near the X-ray emitter in which case that is considerably heavier, possibly about 14 solar masses. A black hole looks an even more likely possibility.

Donald Lynden-
Bell

In summary, the most powerful objects known in nature are the quasars and the radio galaxies. Both of these are related to explosions in the nuclei of galaxies, but probably they are not the furnaces in which most of the helium is formed—the quasars do not appear to be weak in the higher elements. More likely these great objects tap the vaster store of gravitational energy. This leaves us with hydrogen fusion as the source of most starlight, but with Gamow's big bang theory of the origin of most of the helium. His theory has independent support from the detection of cosmic black body radiation about which you can hear more from Dr Sciama in Chapter 5.

Finally, it would be wrong of me to leave you with the impression that astrophysicists can make some sort of sense of all their observations. We are presently baffled by the strange differences in the velocities of similar galaxies that appear close together in the sky. Also, Dr Arp and Professor Geoffrey Burbidge have assembled a rather bewildering array of apparent associations between nearby objects and quasars. They claim that the red-shifts of quasars are not due to the expansion of the universe but, to date, they have not converted the majority of their colleagues. The relevant hard facts known to me are that N galaxies are situated in clusters of galaxies with the same red-shift and one or two low red-shift quasars are, too. Around many low red-shift quasars there are faint galaxies that are candidates for association, but they have not yet been fully investigated.

Until the 1950s, cosmology was a theoreticians' paradise. It was comparatively easy to erect an intellectual edifice and call it a model of the cosmos, simply because there were very few observations to be fitted into the theory. All the model had to do was to account for the expansion of the universe and it was home and dry. Since those days of innocence, the universe has been observed to exhibit several very awkward properties. In the following chapter Dr Dennis Sciama, Fellow of All Souls, Oxford, examines the impact of these observations on one of the most beautiful and intellectually satisfying cosmological models yet devised. This is the steady state model described by Fred Hoyle in a well-remembered series of broadcast talks given in 1950. Whether the steady state model is viable today is a subject of contention in cosmology, and here Dr Sciama gives the background to the debate. L. H. J.

The steady state theory of the expanding universe was proposed in 1948 by Hermann Bondi, Thomas Gold and Fred Hoyle, and according to this theory, the universe has the same large-scale structure at all times. The galaxies still recede from one another in accordance with the Hubble law of red-shifts, but the average spacing between neighbouring galaxies remains the same, so that the general appearance of the universe is unchanged. This becomes possible if new galaxies can form in the spaces left by the old ones, the idea being that they form out of new material that is continually created at just the right rate to keep the population of galaxies steady.

What attitude should we take to such a theory? The regular process of creation which it invokes is incompatible with existing physical ideas, but the creation rate required is far too low for us to be able to contradict it with direct laboratory experiments. The question, therefore, raises deep issues concerning scientific method and the role of physical laws in our understanding of the universe. In the event, the debates on this question have been overtaken by recent developments in astronomy, which strongly suggest that the universe is *not* in a steady state. I shall come to this evidence later, but it seems to me that the issues raised in the debate are sufficiently important to be worth discussing quite apart from this recent evidence.

The most important question is: under what circumstances should we be ready to change the laws of physics as they stand at any given time? This is a rather delicate question. At one extreme we could have people who are ready to introduce a new law of physics to accommodate each new observed fact. At the other extreme, we have the notorious phenomenon of the established scientist who resists to the bitter end the new concepts which experimental facts have forced on to his more open-minded colleagues. Is there a rational middle course between these two extremes?

Ultimately, the only test is the pragmatic one of whose ideas succeed the best. But in the middle of the debate we don't *know* the ultimate outcome and we must be guided by our own sense of the fitness of things. It is interesting that at the present moment in astronomy there are many debates raging of this kind. Do the red-shifts of the quasars demand a new law of physics? Do their high rates of activity? Or, for that matter, the activity in the central nuclei of ordinary galaxies, their gravitational radiation and so on? My own view is that in discussing these localised phenomena, one should work extremely hard to fit them into the *accepted* laws of physics. Only after persistent failure should one introduce new laws; otherwise science loses one of its most important characteristics—its internal discipline.

As I say, this consideration applies to localised phenomena, like quasars and galaxies, but it does seem to me to lose its force when it's applied to the universe as a whole. The point is that the universe is not simply a very large physical system; it's fundamentally *different* in kind from any of its individual parts. The laws of physics as we ordinarily understand them govern the behaviour of any *part* of the universe in a manner that explicitly recognises that other parts also exist. These various parts differ in some respects and are similar in other respects. The similarities we represent by laws of physics, while the differences we call initial or boundary conditions. Let me give a simple example of this distinction. If I throw a ball up into the air, its motion will depend on how hard I throw it and in what direction. But, in so far as air resistance can be neglected, its flight path will for practical purposes be a parabola. So what do we do? Well, we devise laws of motion and of gravity which constrain the flight path of the ball to be a parabola, but we make sure that these laws do *not* constrain the ball's initial velocity or direction.

What applies to the ball applies to larger systems such as the galaxies. The galaxies differ from one another in many respects—in mass, shape, rate of rotation and so on. But in all cases their shapes are related to their motions in accordance with dynamical laws. It's only when we come to consider the whole universe that the situation changes. You see, the point is that there is only one universe. There is no question here of comparing different cases and finding the respects in which they are the same and those in which they differ. So the laws of nature simply cannot apply in the same way to the whole universe as they do to its individual parts. In fact, a strong case can be made out for the view that the structure of the universe is logically *prior to* the laws of nature. On this view, one shouldn't think of the laws of nature as existing independently of the universe (which itself is simply a grandiose example of their application) but rather as a *consequence of* the existence of the universe.

Of course, this view doesn't in itself sanction complete freedom to change the existing laws of nature when one thinks cosmologically. But what it does do is to lay great stress on the total structure of the universe as the most fundamental and logically primitive concept to introduce. As Bondi and Gold argued so cogently in 1948, this means that we should first investigate the simplest of such structures even at the cost of changing, or complicating, the laws of nature. In this spirit, the steady state model was introduced because its structure is so much simpler than models which begin with a big bang of infinite density and then evolve with time.

I am anxious to stress this approach to the steady state model because many people believe that the main reason it was introduced was the time-scale problem that existed in 1948. The age of the universe since the big bang was *then* believed to be 2,000 million years, while the age of the galaxy appeared to be some five times greater. This discrepancy would be automatically removed by the steady state theory, because in that theory, galaxies would have existed at all times, so any particular galaxy, like our own, could certainly be as old as 10,000 million years. However, as Dr Peach points out in Chapter 2, astronomers revised the extragalactic distance scale in the early fifties, and thereby increased their estimate of the age of the universe to the point where the time-scale difficulty disappeared. For many people, this development removed whatever attractions the steady state model may have had for them. I believe this was a profound mistake. While the relation of the steady state model to observation is of vital importance, the reason for introducing the model in the first place was not so much an empirical but a logical one. Disagreement with observation would serve to disprove the steady-state model, but the agreement of other models with observation is irrelevant.

Of course, there was no disagreement in 1948. As a matter of fact, this came as a surprise to Bondi, Gold and Hoyle. Attracted by the structural simplicity of the steady state model, they at first assumed that there must be some decisive evidence against it, to account for the fact that astronomers never seemed to consider it. Investigation showed that in 1948 there was no such evidence and that the silence of astronomers on the subject of a steady-state universe arose rather from their habit of following physicists in placing the laws of nature above the structure of the universe. They just took for granted that the expansion of the universe meant that it had a structure which was changing with time.

Encouraged by the absence of contrary evidence, Bondi, Gold and Hoyle thereupon made their famous proposal. In retrospect, it's not surprising that they could, as it were, get away with it. Observations of galaxies were then confined to distances that were small compared with the radius of the universe. Equally, the time in the past when the observed radiation was emitted was small compared with the age of the universe. We simply hadn't probed far enough into the universe in either space or time to decide whether, for instance, galaxies seen as they were in the past were closer together on average than they are today. Another test would have been to decide whether galaxies have the large spread of ages that would be expected on the steady state model. Even today, we cannot date the galaxies with sufficient confidence. So, steady state cosmology flourished through lack of observational evidence to the contrary.

There remained to work out the various consequences of the theory and to try and frame physical laws which would be compatible with the continual creation of matter. The most important success of this period was the theory of the formation of the elements, worked out originally by Hoyle and later also by William Fowler and by Margaret and Geoffrey Burbidge. The need for this theory arose because, by doing away with the big bang high density origin of the universe, the steady state model also did away with the nuclear reactions occurring at the epoch of high density. George Gamow had claimed that it was precisely these nuclear reactions which had led to the formation of all the elements heavier than hydrogen which we now find in the universe. Clearly, the steady state theorists had to rely on events occurring *now*, in galaxies, to provide regions hot enough for the necessary nuclear reactions to take place. They were led to concentrate on supernova explosions, and in the hands of the present generation of nuclear astrophysicists, this theory of the formation of the elements appears to be very successful. There is one important exception and that is the element helium. It's likely that 10 per cent of all the atoms in our galaxy are helium atoms, whereas the known processes occurring in the galaxy seem able to convert only one per cent of the hydrogen into helium. This discrepancy will form a central topic of my discussion of the big bang later in this chapter.

Although the supernova theory of the origin of the elements seemed very promising, the tide began to turn against the steady state model with the revolution brought about by radio astronomy. The first hint of this came in 1955 when Martin Ryle and Peter Scheuer created a considerable stir by publishing the first counts of radio sources. The early counts were, in fact, inaccurate, and the lack of optical identifications for most of the radio sources concerned enabled the ardent steady state theorist to doubt the conclusion that the intrinsic properties of the radio sources were changing with cosmic epoch, a conclusion clearly at variance with the steady state model. The resulting controversy, mainly between Hoyle and Ryle, was animated and resourceful and echoes of it still resound today.

Well, who was right? The majority of radio astronomers probably side with Ryle, although an obstinate minority feel that the last word has still to be said. My own view, for what it's worth, is that the argument from the radio source counts is rather impressive, but that if it were the only evidence against the steady state model, I would be reluctant to abandon my attachment to that beautiful theory.

Unfortunately, it is *not* the only evidence. Hard on the heels of the radio source counts came the discovery that amongst these radio sources were the now notorious quasars (plate 5). If the red-shifts of quasars are due to the expansion of the universe, these objects are so bright optically that they can be observed by large telescopes at distances comparable with the radius of the universe. This would make them ideal objects for exploring the large-scale structure of the universe. For this reason I started in 1966 to study the numbers of quasars at different red-shifts in order to compare the observations with the predictions of the steady state theory. At first, the comparison seemed to be working out favourably until my student, Martin Rees, pointed out to me that there were far too many quasars of large red-shift to be compatible with the steady state model. My own disillusionment with the model dates from that time, and a more careful analysis by Maarten Schmidt, the discoverer of the quasars, only served to confirm Rees's point. There is, however, a let-out. Several astronomers, including Fred Hoyle and Geoffrey Burbidge, have argued that the red-shifts of quasars may have nothing to do with the expansion of the universe. In that case, the relative number of different red-shifts would be irrelevant to the cosmological problem.

Again, the majority view is not sympathetic to this idea, and again I must say that if this were the only evidence, I don't think I would have abandoned the steady state model. But, taken together with the radio source counts, the evidence is clearly beginning to mount up.

I now come to what for many people is the most compelling evidence of all. It's also, perhaps, the most magical of all astronomical discoveries. Indeed, the whole process of initial accidental discovery, subsequent interpretation and final recollection that it had all been predicted twenty years earlier, seems now just like a fairy story. The story starts in 1965, when two radio astronomers at the Bell Telephone Laboratories, Arno Penzias and Robert Wilson, were observing the sky at a wavelength of 7 centimetres with a horn antenna. Their observations were noisier than they had expected and they tried to eliminate the noise by readjusting the antenna. All their attempts failed. They then reluctantly concluded that the noise was *not* generated in their detector but represented, in fact, a signal coming in from outer space, with approximately the same intensity in all directions. This intensity was very high—about one hundred times greater than would be expected to arise from all the known types of radio sources in the sky.

An interpretation of this phenomenon was immediately proposed by Robert Dicke and his colleagues, who were working a few miles away at Princeton University. As luck would have it, Dicke and his group were at the time engaged in constructing a receiver to work at the nearby wavelength of 3 centimetres, in the precise hope of detecting an excess signal from the sky far stronger than that due to known radio sources. Dicke's reasoning was as follows: if the universe has expanded from an early dense phase, this phase may itself have been preceded by a collapse from a still earlier dilute phase. During the collapse, the Doppler effect which, in our present universe, gives rise to a red-shift, would be reversed and any radiation produced would have been amplified by a blue-shift. Now, shift the spectrum of any radiation to the blue, and you—in effect—raise its associated temperature, so the state of maximum density could have been very hot. This heat would have survived the bounce into the expansion phase and so the big bang would have been a hot big bang.

There is an important consequence of this idea. In the high density phase, the heat radiation would have interacted strongly with matter and would rapidly have come into what the physicist calls thermal equilibrium. This means that the spectrum of the radiation would have taken on a universal character. The energy present in the radiation at different wavelengths would then have been determined just by the temperature of the system, in accordance with the so-called black body law associated with the name of Max Planck. Moreover, once this black body spectrum was set up, it would be maintained throughout the expansion of the universe, even when the matter in it became so dilute that interaction with the radiation became unimportant. All that would happen would be a cooling of the radiation as the universe expands. Dicke estimated that the temperature of the universe was a thousand million degrees just one hundred seconds after the big bang, and that now, 10,000 million years later, the radiation would have cooled down to the apparently low tempertaure of about 3 degrees absolute—that is, about minus 270 degrees centigrade.

Dicke's point was that this radiation would even now still have a black body spectrum, that at a few degrees absolute the spectrum would peak at radio wavelengths, and that this peak would be much stronger than the background radio noise due to individual radio sources like radio galaxies, quasars, or even our own galaxy. He accordingly set out to detect this excess radio noise, but before his equipment was completed, there came the announcement from Penzias and Wilson of their puzzling discovery of excess radio noise. Naturally, Dicke and his colleagues immediately proposed that Penzias and Wilson had detected black body radiation left over from the hot big bang. This detection referred to a measurement at one wavelength only, namely 7 centimetres. If the true spectrum were that of a black body, then that measurement implied that its present temperature would be close to 3 degrees absolute. Clearly the test of Dicke's hypothesis would be to measure the background at different wavelengths, to see whether the spectrum really was that of a black body at 3 degrees absolute.

Not surprisingly, the first such measurement came from Dicke's group, who found that the intensity at 3 centimetres was in agreement with the black body hypothesis. Since then, many different groups have reported results for wavelengths ranging from 70 centimetres to 3 millimetres. The long-wave limit is set by the radiation from our own galaxy, which begins to dominate at that point. The short-wave limit is set by absorption in the earth's atmosphere. Over this range of wavelengths, the agreement with a black body spectrum is good. But in all fairness, I must point out that these measurements are difficult and one needs to beware of a band wagon effect that can occur once a result is expected.

And there's another thing: throughout most of this range

of wavelengths, the black body spectrum has a particularly simple form in which the intensity is proportional to the square of the freuqency. One could envisage a number of processes that give such a spectrum over a limited range of wavelengths. The really characteristic part of the black body spectrum at 3 degrees absolute occurs around 1 millimetre, where the spectrum peaks and then falls off rapidly at shorter wavelengths. This decisive part of the spectrum is, unfortunately, sealed off by the opacity of the earth's atmosphere. There are indirect measurements available in this wavelength region from molecular effects in interstellar space, and these are in good agreement with the black body hypothesis. There have also been a few rocket and balloon flights above the main part of the earth's atmosphere, but these difficult measurements are at the moment in conflict with one another. No doubt we shall have to wait for a satellite before we obtain the really decisive evidence. Meanwhile, the general consensus of opinion is that this radio background probably does have a black body spectrum.

Well, where does this leave the steady state theory? The essential point is that the matter in the universe *today* is far too dilute to bring any excess radiation into thermal equilibrium in the time available. So while we could accept that radiation may be produced along with the continual creation of matter, this radiation could *not* be brought into equilibrium. It would be stretching our credulity too far to assume that when this radiation is created, it already *has* a black body spectrum. There have been attempts to attribute the observed background to a new class of radio source, but I do not think these attempts have been at all successful. No, I'm afraid there seems to be no escape. Taken together with the evidence from the radio source counts *and* the quasar redshifts, the excess background of radiation creates very grave difficulties for the steady state theory. The 3 degrees absolute background radiation implies that processes must have occurred to bring the radiation into thermal equilibrium and no one has been able to find a rapid enough process in the universe as it exists today and as it always would have existed in a steady state universe. But this problem would be solved if the density of matter in the universe were much greater in the past than it is now. Well, we already know from the red-shifts of the galaxy that the universe is expanding, so we would, in fact, expect the universe to have been denser in the past, so long as we confine ourselves to conventional theories in which matter is not continually created. The simplest of such theories then lead to the conclusion that the universe had an infinite density at a finite time in the past, this time being about 10,000 million years ago. Here then is a possible origin for the universe. The three degrees radiation that we observe today suggests that this beginning was hot, the idea being that the radiation cools down as the universe expands. In fact, Einstein's general theory of relativity enables us to calculate the temperature of the universe in its early stages in terms of the time as measured from the big bang itself. This relation happens to be a very simple one: the temperature multiplied by the square root of the time is a constant. If we measure the temperature in degrees absolute and the time in seconds this constant is 10,000 million (10^{10}); that means that one second after the big bang the temperature was 10,000 million (10^{10}) degrees absolute and that after one hundred seconds it had dropped to 1,000 million (10^9) degrees. This simple relation does not hold exactly in recent stages of the expansion but it is not a bad approximation for rough purposes. Taking for the present age of the universe 10,000 million years we would find for the present temperature about 15 degrees which is of the same general order as the 3 degrees actually observed. It would, of course, be desirable to check these ideas in an independent way. Is there any other evidence that the universe once had a temperature as high as 1,000 million degrees? Well it turns out that there is. As Gamow first pointed out in the 1940s, it's important to examine what nuclear reactions would occur in the high density, high temperature phase of the universe. Gamow's calculations have been much improved in recent years, but he was able to show in a general way that a significant amount of helium would be built up out of hydrogen by these early reactions. This led Gamow to introduce the idea of the hot big bang and he predicted that the present temperature of the radiation would lie in the general vicinity of ten degrees absolute.

However, in the mid-forties radio astronomy was in its infancy and Gamow didn't realise that his predicted radiation field would be detectable, and that, indeed, at centimetre wavelengths it would easily dominate all other sources of extraterrestrial radiation. The subsequent success of the supernova theory of the origin of the elements led to Gamow's prediction being forgotten. Dicke and his colleagues reintroduced the hot big bang in 1965 in ignorance of Gamow's work, and Penzias and Wilson discovered the excess background radiation at 7 centimetres by accident.

Modern calculations of the nuclear reactions occurring in the early stages of the universe show that the crucial period occurs about one hundred seconds after the big bang when the temperature was 1,000 million degrees. At that time an appreciable amount of helium would have been built up out of hydrogen but very little of the heavier elements. By contrast the supernova calculations show that while the observed abundance of the heavier elements might well be formed in stellar explosions, the amount of helium formed in these explosions would be much less than in the hot big bang. We, therefore, have here a clear-cut opportunity of testing the whole body of our cosmological ideas. Is the observed abundance of helium in the galaxy the same as we would calculate from the nuclear reactions occurring in the hot big bang?

The provisional answer to this question, as Professor Lynden-Bell points out in the previous chapter, must be that the hot big bang passes this test remarkably well. I say provisional because the observations of the helium abundance are difficult to make and there are some uncertainties of interpretation. But by and large the agreement is very good. There are, in fact, several different ways of measuring helium abundances and the striking thing is that apart from one or two discrepancies they all agree that about 10 per cent of all the atoms in the galaxy are helium atoms. Even the few discrepancies seem less worrying now than they did originally. The observations range from helium lines in the optical spectra of stars and radio lines produced in clouds of interstellar gas to cosmic rays coming from the sun and indirect arguments concerning theoretical models of stellar structure. Nearly all agree on a helium abundance of 10 per cent and this is precisely what would be expected if the helium were formed in the hot big bang. The resulting relation between the 3 degree background, the helium abundance and the hot big bang represents the most important advance in cosmology since the discovery that the universe is expanding.

It's not the only advance, however. The 3 degree background makes its presence felt in other ways as well. It does so in an active sense in that it is important dynamically and it does so in a passive sense in that the degree to which it has the same intensity in different directions tells us a number of important things about the structure of the universe.

Let us begin with its dynamical role. It might seem at first sight that the temperature of the black body radiation is very low—only three degrees above absolute zero, the sort of temperature we associate in the laboratory with liquid helium. Nevertheless, from an astrophysical point of view three degrees absolute is a very high temperature. A radiation field at this temperature distributed throughout the galaxy would have about the same energy per unit volume as the other known modes of interstellar excitation—starlight, cosmic rays, magnetic fields and turbulent gas clouds. So even in our galaxy the cosmological background radiation would be for many purposes as important as the well-known energy modes of local origin. But in the space between the galaxies these localised energy densities probably drop off by a factor of at least a hundred, whereas the black body component would maintain its energy density unchanged.

Under these circumstances we might expect this component to exert an observable influence on astrophysical processes, especially on those involving high energy particles like cosmic rays that can interact with the black body photons. The most important of these interactions involves the electrons in the cosmic rays. These fast electrons are believed to give rise to the radio emission of our galaxy as they spiral around the interstellar magnetic field. The same process is probably responsible for the radio emission of other normal galaxies, radio galaxies and quasars. Now these electrons interact with the low energy photons of the 3 degree background converting them into X-rays and losing energy in the process. It has not yet been possible to observe this interaction directly so we can't claim to have confirmed the universal character of the 3 degree background from these considerations. However, it is important to note that the interaction processes depend sensitively on the temperature assumed for the black body radiation. For instance, a temperature as high as 10 degrees would produce large interaction effects that couldn't possibly have escaped detection, so such a high temperature could confidently be ruled out. By the same token the interaction effects occurring in the past when the radiation field was hotter than it is today would also have been more important. It is quite possible that the resulting X-rays make the main contribution to the X-ray background that is observed to come from outer space. However, there are other explanations of this background and this question remains unsettled. Nonetheless, the resulting energy drain on the electrons almost certainly plays an important part in the evolution of radio galaxies and quasars.

I now come to the passive role of the 3 degree background: its use as a probe for investigating the universe. Here its most important property is its isotropy, that is the degree to which it has the same intensity in different directions. Careful searches have failed to reveal any significant variation with direction, the precision achieved being about one part in a thousand. This is by far the most accurate measurement in cosmology and it tells us a great deal about the

E

universe. The first thing it tells us is that the earth must be moving at a very low speed relative to the universe as a whole; otherwise the Doppler effect would be important and the radiation field would be detectably hotter in the direction of the earth's motion and cooler in the opposite direction. The observed absence of such an effect to a precision of one part in a thousand limits the speed of the earth to 300 kilometres per second. Two observers have, in fact, claimed to detect a positive effect at just this level but their claims have not been generally accepted and only the upper limit is regarded as well established.

What is the significance of this result? In the first place we must consider what it means to talk about this speed. Relative to what is it being measured? The answer is that it is relative to the 3 degree background and since that background comes to us from the distant reaches of the universe it is relative to the universe as a whole. It is important to realise that this conclusion is not in conflict with the special theory of relativity which asserts that absolute velocity has no significance. The point is that the black body radiation connects us to the whole universe which, therefore, comprises one physical system. Special relativity does not, of course, state that relative velocities within a physical system cannot be measured. What it does say is that the motion of a complete system, in this case the whole universe, with constant velocity would have no observable consequences. The upper limit of 300 kilometres per second is of great astronomical importance because the motion of the earth under various localised forces would be expected to be of this order. The earth goes round the sun at only 30 kilometres per second, but the sun goes round the centre of the galaxy at about 250 kilometres per second. Moreover, the galaxy moves relative to the other galaxies of the local group at about 100 kilometres per second, and the local group may move as a whole under the gravitational influence of neighbouring clusters of galaxies. Clearly the fact that the resultant of all these motions is less than 300 kilometres per second tells us a good deal about the hierarchy of irregularities that are acting on the earth.

There is a further point here. One of the deepest problems of dynamics is the relation between what are known as non-rotating frames of reference and the structure of the universe. A dynamically non-rotating frame is defined as one in which no centrifugal forces have to be introduced to ensure that Newton's Laws of Motion hold good. According to the so-called Mach Principle (which I describe more fully in Chapter 8) such non-rotating frames are determined by the universe as a whole rather than by absolute space. This means that the universe itself should not appear to rotate if viewed from a dynamically non-rotating frame. How accurately is this requirement satisfied in practice? Before the measurements on the 3 degree background, the best one could say was that the rotation period of the universe could not be less than 10,000 million (10^{10}) years. So the universe could have gone once round since our galaxy was formed. However, a rotation with this limiting period would lead to a motion of the earth through the 3 degree background which would have been easily observed. In fact, one can now say that any rotation of the universe must be at least 10,000 times slower. So since our galaxy was formed the universe couldn't have gone round more than one ten thousandth of a turn. This strongly suggests that the Mach Principle is correct, and that local dynamics can only be understood in a cosmological context.

I now turn to other consequences of the high isotropy of the 3 degree background. Since the intensity of the radiation decreases as the universe expands the isotropy must mean that to high precision the universe expands at the same rate in all directions, and that different parts of the universe all have the same starting temperature. This uniformity in the large-scale properties of the universe is a godsend to the mathematician, who is faced with the problem of finding cosmological solutions to Einstein's equations of general relativity. These equations are extremely complicated and one can find solutions only if one assumes that the universe is highly symmetrical. Now it seems that this symmetry assumption is a very realistic one.

However, what is a godsend to the mathematician is a headache to the astrophysicist. He wants to know why the universe is so isotropic. Is it a matter of initial conditions in the hot big bang which cannot be further explored, or do there exist smoothing processes which remove any initial anisotropies as the universe evolves? This question is under active investigation at the moment and it is not at all clear what the answer is going to be.

A similar problem concerns the homogeneity of the universe, that is, the extent to which it is the same in different places taken at the same cosmic time. We know that there are galaxies, clusters of galaxies and possibly super-clusters, although we do not understand in detail how they came to be formed. Are there any larger-scale irregularities, or can we say that averaged over several clusters of galaxies the universe is the same at all places taken at the same time? The isotropy of the 3 degree background gives us some information on this. For if there were large-scale irregularities which differed somewhat in detail along different lines of sight then their gravitational influence on the 3 degree background would induce variations in its intensity with direction. But variations like these are not observed, and this tells us that on a large scale the universe is homogeneous as well as isotropic. Again, a godsend for the mathematician and a headache for the astrophysicist.

I want to end with what is perhaps the most surprising inference we can draw from the isotropy of the 3 degree background. This concerns the question: did the universe really begin not just with a big bang, but in a singularity where all the known laws of physics break down in a state of infinite density? It used to be argued at one time that this singularity was an artificial consequence of the mathematical assumption that the universe is exactly isotropic. If one assumes that every particle in the universe is moving away from us exactly radially it is no surprise that all these particles converge on us at precisely the same moment in the past. Introduce some realistic irregular transverse motions, it was argued, and the singularity would go away. We now know that this is not true. Thanks to some powerful theorems due to Penrose, Hawking and Ellis we know that in a wide variety of circumstances the effects of self-gravitation are so great in general relativity that it is very difficult to avoid a singularity. In particular the high isotropy of the 3 degree background can be used to demonstrate that the actual universe is sufficiently close to an exactly isotropic one for the singularity theorems to apply.

So the 3 degree background radiation enables me to answer the question 'how did the universe begin?' If general relativity is correct the universe began with a very big bang indeed.

The steady state theory still has a few adherents who rebel against the wholesale acceptance of the big bang which, by its very nature, assumes that there are aspects of the universe that can never be known. Little is understood about the big bang itself except that all information is destroyed as one approaches nearer and nearer to it, and that we can never know what preceded it or caused it—if, indeed, such phrases can have any meaning at all. By contrast the orderly progression of the steady state theory, in which the large-scale aspects of the universe have always been and always will be the same, still has a strong, aesthetic appeal.

One who has done much to shore up the structure of the steady state theory against the battering rams of awkward observations is Jayant V. Narlikar, formerly of the Institute of Theoretical Astronomy, Cambridge, and now Professor of Astrophysics at the Tata Institute of Fundamental Research, Bombay. L. H. J.

STEADY STATE DEFENDED

Professor J. V. Narlikar

Right from its birth in 1948 the steady state theory has played an important part in the development of cosmology. Even its opponents, and there are many, will not deny this fact. On the theoretical front the radical ideas underlying the theory have provoked controversies. On the observational front the theory has been constantly under attack, but this is only to be expected. After all, unlike its rivals the big bang cosmologies, the steady state theory makes clearcut predictions which can be tested by observations. If, as Bondi suggests in the opening chapter of this volume, one criterion for a good scientific theory is that it should 'live dangerously', then this alone would give the steady state theory a flying start over its competitors!

However, for a scientific theory to live dangerously is one thing. But for the theory to die an untimely death in consequence is quite another. Have astronomical observations really killed the steady state theory? If this question were asked in an opinion poll of astronomers, a majority would probably answer it in the affirmative. Yet, history of science shows that theories are not discarded on a majority vote but only when the observations are unambiguously against them. Do the observational claims against the steady state stand up to an objective assessment of the data? I will examine this question in the central part of this chapter.

First, I would like to begin the discussion by describing what the theory is about, because considerable confusion results from time to time when the observers try to interpret what the steady state theory in fact has to say. Even its authors differ among themselves as to the proper approach to the theory. The approach of Bondi and Gold is a deductive one,

Chapter 6

starting from one central principle. Hoyle's approach makes use of equations which are a modified form of those which describe Einstein's general theory of relativity.

Bondi and Gold argued as follows. In cosmology, we study the structure of the universe by looking at its distant parts and comparing them with our own neighbourhood. These observations are carried out by using electromagnetic waves, e.g. optical, radio, X-rays, etc., all of which travel with the speed of light. Therefore, when looking at the distant parts of the universe we see them as they were a long time ago. Any meaningful comparison of these distant parts with the ones nearby presupposes that the laws of physics are the same in the two cases. Since the universe, by definition, includes the study of *all* observable phenomena, we may expect the laws of physics also to be somehow determined by the universe as a whole. If the state of the universe was once very much different from what it is now, what guarantee do we have that the laws of physics were the same in the past as they are now? Indeed, the only way one can definitely guarantee that the laws have always been the same is by postulating that the universe has likewise always been the same. Bondi and Gold formulated this point of view in the form of the so called 'Perfect Cosmological Principle', which states that the universe, in the large, is unchanging in time as well as space.

A remark on the word 'perfect'. Even before Bondi and Gold, cosmologists did use a 'cosmological principle' in a restricted sense. This principle assumed large-scale homogeneity and isotropy of the universe at *any given epoch*. This meant that two observers at different locations in the universe but at the same epoch of time would see the same large-scale view of the universe in all directions. By adding the qualification 'perfect', Bondi and Gold wished to emphasise the large-scale regularity of the universe not only over space but also over time.

From the perfect cosmological principle Bondi and Gold deduced that the universe must always expand. The only other possibilities are that it is either static or always contracting. In either case, as Bondi himself shows in Chapter 1, we run into the Olbers paradox, with a very high or infinitely bright sky background. Only if the universe expands continually will the night sky remain dark. Incidentally, this is an example of the deductive power of the perfect cosmological principle, when taken together with the information available from our immediate neighbourhood.

One immediate deduction from this expansion of the universe is that new matter must be created continually.

The injection of new matter serves to keep the density of matter constant, as required by the 'steady state'. This is in contrast to the big bang models where matter is created once and for all in a single event. The steady state creation rate is extremely small, about one gramme of matter appearing in a volume of one litre in about a trillion trillion years. (Trillion = million million million.)

The reasoning leading to the perfect cosmological principle, and the deductive power of the principle are the attractive features of the Bondi-Gold approach. Nevertheless this approach has certain drawbacks. For example, it was never stated precisely over how large a scale of space and time the perfect cosmological principle applies. Clearly it must apply to some average condition of the universe, and a certain amount of fluctuation about this condition is to be expected. If an observer does not allow for this in the interpretation of his data, he is likely to conclude that he has found evidence against the theory when in fact he is merely observing a fluctuation from the average steady state. The lack of specification of the degree of departure from the steady state condition has thus led to confusion among those who want to interpret the predictions of the theory. Also, the theory lacks field equations which could tell us, as in the case of big bang models, how to relate quantitatively the various observable characteristics of the universe. There are no physical laws to tell where the new matter comes from. All we can say is that it is one of the consequences of the perfect cosmological principle.

Hoyle approached the steady state theory from another direction. He argued that a cosmological theory must answer the question: 'Where did the matter we now see around us come from?' This question is by-passed in big bang cosmologies, in which Einstein's equations describe the behaviour of matter only *after* it has been created. To study the actual creation process, Hoyle modified the Einstein equations by adding further terms to them. These terms provide an energy reservoir from which matter can be created. At first sight such an attempt might appear doomed from the outset. Any energy reservoir is expected to be depleted gradually for two reasons: firstly the fact that matter is being extracted from it means it loses an equivalent amount of energy according to the Einstein rule $E = mc^2$, secondly the expansion of the universe dilutes any reservoir of energy.

How can a steady state be maintained in spite of this? Hoyle got round this problem by an ingenious method. He made the cosmological reservoir of *negative* energy. The strength of such a reservoir *increases* by matter creation because extraction of positive energy from a negative energy reservoir makes the original energy density even more negative. The two processes described above now work in the opposite sense and they are held in balance in the steady state universe. Departure from this balance can lead to fluctuations from the steady state picture, but now we have quantitative estimates about the extent of such fluctuations.

Of course, criticisms can be levelled against this approach. The concept of negative energy is unusual in physics, although it is worth pointing out that gravitation also has negative energy. On a local scale negative energy reservoirs present mathematical problems, especially in quantum physics. However, a preliminary investigation has shown that these difficulties are unlikely to be present when the reservoir is spread over the whole universe. A second, and more important criticism is that if we set out to modify Einstein's equations we can do so in a large number of ways —whereas a good physical theory should have as little arbitrariness as possible. To a great extent this criticism was met when Hoyle adopted the starting point suggested by Maurice Pryce in 1960. This introduces the simplest possible modification, where the energy source is a 'massless scalar field'. Even this field is not required in a more recent formulation given by Hoyle and Narlikar.

Having described the basic ideas underlying the steady state theory I can now examine how the theory has faced the many challenges placed in its way by observational astronomy. The first challenge came from the observed abundances of chemical elements. The work of Gamow and others had shown that heavy nuclei could possibly be synthesised from neutrons and protons in the early stages of a big bang universe. In the first few moments after the big bang the universe has a high temperature and density of radiation and matter, and these are indeed conditions suitable for nucleosynthesis. It was argued that such conditions are never present in a steady state universe which therefore fails to account for the presence of heavy elements. However, it was soon realised that the wide range of conditions operating inside the stars in their various evolutionary stages also permit the synthesis of all the observed elements. The pioneering work of Geoffrey and Margaret Burbidge, William Fowler and Fred Hoyle demonstrated in great detail how this is achieved. Indeed the tables were turned back on the big bang cosmologies when it became clear that the original expectation, that the nucleosynthesis could be done entirely in the big bang, is not correct. At best we can get hydrogen and helium from the hot big bang, but not the heavier elements like carbon, oxygen, etc. For these we need the stars anyway! Recent calculations with high-speed computers have confirmed these conclusions.

The next challenge came from the Stebbins-Whitford effect. It is known that because of cosmological red-shift (the fractional increase in the wavelength of light from a distant source, when received by a terrestrial observer) the distant galaxies would appear redder than the nearby ones. Stebbins and Whitford found reddening in distant galaxies over and above that arising from the red-shift. This meant that galaxies at earlier epochs were systematically redder than they are now, thus implying a departure from the steady state situation. Subsequent work showed, however, that the effect was a spurious one, and there is no excess reddening.

I now come to two cosmological tests which have frequently claimed to have disproved the steady state theory. The first is the extension of Hubble's velocity distance relation and the second is the counting of radio sources.

Hubble's original observations, now nearly four decades old, showed a linear relation between the red-shift and the distance for nearby galaxies. These observations have since been extended to more and more distant galaxies, the most extensive work being due to Sandage and his collaborators with the 200-inch telescope at Mount Palomar. The main purpose of these observations is to measure accurately the rate of expansion of the universe and to try to distinguish between the different cosmological models. John Peach has given details of this work in Chapter 2.

The observations are plotted on the so-called Hubble diagram. On the x-axis we have the apparent magnitude and on the y-axis the red-shift of the galaxy. Apparent magnitude measures the faintness of the galaxy as seen from the terrestrial observatory. If all galaxies were equally bright intrinsically, the more distant they are the fainter will they appear. Thus we have a crude measure of distance along the x-axis. In Fig. 7 we have some theoretical curves predicted by a range of big bang models and the steady state model.

If we have a large number of galaxies with red-shifts, say in excess of 0.2, we may be able to tell which curve best represents the data. For red-shift values under 0.2, there is no substantial difference between the predictions of different cosmological models. And here we encounter a frustrating situation. Our measures of distance get less and less reliable for more and more distant galaxies. For red-shift values under 0.2, a fairly reliable measure can be obtained, as Peach described, by looking at only the brightest members of clusters of galaxies. From a study of nearby clusters it can be ascertained that the variation of intrinsic brightness among this class of galaxies is not too great. However, if we now look for similar galaxies with red-shift values over 0.2, we find only three so far, with red-shift values 0.29, 0.36 and 0.46! The last of these was discovered with the help of radioastronomy in 1960. This red-shift belongs to a cluster of galaxies associated with the radio source 3C-295. It was hoped that more high red-shift galaxies might be discovered in this way by identification with radio sources. Although a number of identifications have been obtained, none have the high red-shifts we are looking for. So the conclusions depend essentially on three points!

A few years ago it used to be stated that the model best favoured by the data is one where the expansion of the universe is slowing down so that it will eventually contract. However, recent examination of the data indicates various ifs and buts associated with distance measurements, and the range of uncertainty is such as to permit a wide range of curves of Fig. 7, including the steady state curve.

At this stage we may ask 'What about the quasi stellar objects with their large red-shifts?' Red-shifts of the order of 2 are very common among the quasars. The discovery of early quasars raised the hope that they would settle this question one way or the other, and indeed the first few quasars did seem to lie on a well-defined line. This state of affairs was not to last for long however. In 1966, Hoyle and Burbidge pointed out that the quasars whose red-shifts were known then did not lie on any well-defined line but had a huge scatter. This conclusion has not been changed since the discovery of more quasars. A typical Hubble diagram for quasars (Fig. 8) looks a scatter diagram and only an eye of faith might spot a smooth curve running through the points. Indeed, one is led to suspect that because the quasars do not seem to follow the Hubble law, their red-shifts may not be due entirely to the expansion of the universe. In spite of such evidence, this conclusion has been resisted by many astronomers.

Perhaps the most controversial attack on the steady state theory comes from radioastronomy. This involves the counting of radio sources down to specified flux density. (Flux density is a measure of the amount of radiation received over unit area in unit time in a given frequency range.) The basic idea of this test is quite simple. Suppose we live in a simple universe in which Euclid's geometry holds, there is a uniform distribution of sources emitting radio waves, and all sources have equal strength. If we draw a sphere around us, the faintest looking sources in it will be

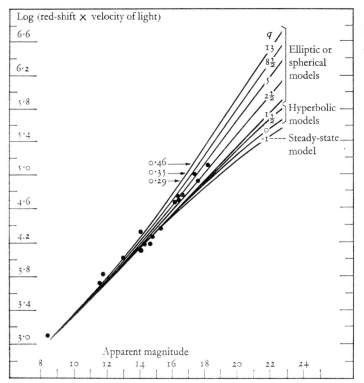

Fig. 7: Data by Sandage and Baum plotted as logarithm of red-shift against apparent magnitude.

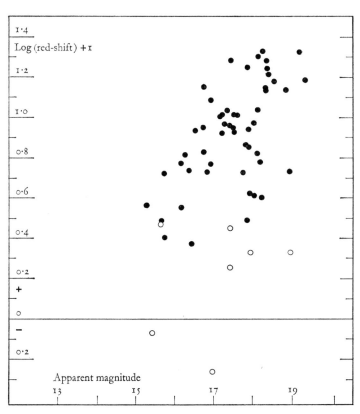

Fig. 8: Hubble diagram for quasars

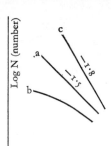

a: Euclidean universe
b: Steady-state universe
c: Observed curve
 (3C-survey)

Fig. 9: Graphs of radio source flux levels against number observed N.

those on the boundary. The flux level of such sources will be inversely proportional to the square of their distance from us. On the other hand, if we count all the sources brighter than these on the boundary, they will be inside the sphere. Their number will be proportional to the cube of the radius.

In an expanding universe the above assumptions may have to be changed. If we keep the distribution of sources emitting radio waves and their strength unchanged but give up Euclid's geometry, we can easily work out the log N (number) log S (flux density) curve for any given cosmological model. Fig. 9 shows three curves. Curve a is the straight line with slope -1.5. Curve b is that calculated for the steady state universe. Curve c, which is the observed curve, has slope -1.8. The relative spacing of the curves above or below one another is not important; only their slopes are.

The observations have had a somewhat checkered history. Early Cambridge surveys by Ryle and his collaborators indicated a slope of -3 for the log N—log S curve, in violent disagreement with the curves a and b. Similar surveys of the southern hemisphere by Mills and his colleagues indicated a less steep slope, around -1.8. Later Cambridge surveys also settled down to this value. The curve c shown is the 3C-curve (the third Cambridge survey). Further Cambridge surveys, the 4C and 5C, continue this curve to lower flux values. These surveys indicate that the curve begins to flatten at lower flux values, much as one would expect from a continuation of the steady state curve b. So the real controversy lies about source counts at high flux levels.

In the last few years more detailed work has been done at this end of the log N—log S curve. We have already seen that on the Hubble diagram the quasars behave in a peculiar fashion. They differ from galaxies in many ways and it seems somewhat artificial to count them together with galaxies. The revised 3C-catalogue can therefore be analysed in the following way. First we single out those sources which can be identified with objects seen through an optical telescope. Of these, those which are identified with galaxies are placed in one category. Those identified with quasars are placed in another. The unidentified sources form the third category.

The slope of log N—log S curve for the sources in the first category is not significantly different from -1.5. However, a plot of red-shift against radio brightness for these radiogalaxies reveals no correlation at all. This means, the radiogalaxies differ widely in their intrinsic brightness and the assumption that all sources have equal strength made earlier is not valid. Thus we cannot use this log N—log S curve to prove or disprove cosmological models.

A similar situation prevails for the quasar radio sources. But, in this case the red-shifts are large. If the red-shifts are of cosmological origin, we could ascribe their log N—log S curve to a powerful evolutionary effect that contradicts the steady state theory. However, considerable doubt exists about the nature of quasars' red-shifts, as we shall see later.

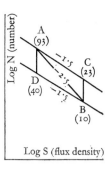

Fig. 10: Unidentified sources.

The unidentified sources appear to hold the key to the whole question. Ryle and his group have argued that these sources are not nearby and their log N—log S curve which has a steep slope of —2.5 or so, indicates an evolutionary universe. Hoyle argues that they are comparatively nearby and the steep slope is not cosmologically significant. Hoyle's argument is illustrated briefly in Fig. 10. The line AB has slope —2.5 between flux levels .5 and 12.5, the 3C values at the frequency 408 MHz. At the high flux end B we have ten sources, and at low flux end ninety-three sources. The lines AC and BD have slopes —1.5. If the universe were Euclidean and we accept the point B as correct, we find that we have an observed excess of fifty-three sources at five flux units. If the point A is correct, we have an observed deficit of some thirteen sources at the high flux end. Since the survey is over three steradians in the sky, the deficit in this latter case amounts to only of four to five sources per steradian! This can be easily accommodated in normal statistical fluctuation. The inconclusive nature of the present log N—log S data as far as cosmology is concerned, was emphasised by Kellermann in 1971 in the Warner lecture given to the American Astronomical Society.

These issues can be resolved if we are able to get red-shifts of these sources. They are mostly believed to be radio galaxies. If the red-shifts turn out to be large, Ryle will be proved right. If the red-shifts are not significantly larger than those for the sources in the first category, then we must agree with Hoyle that the test does not have any cosmological value. Another possible check on this issue would be the distribution of these sources over the sky. If the distribution shows anisotropy, then they cannot be very far away.

Clearly the quasar sources play an important part in the cosmological question, if it can be established that their red-shifts arise from the expansion of the universe. We have already seen reasons to doubt this. Recent work has revealed further evidence to support these doubts. For instance, Harlton Arp at the Hale Observatories has found cases where quasars appear to be visibly connected to galaxies—but where the red-shifts of the connected objects are quite different. There are quasars with multiple red-shifts posing further problems to the theoretician. At one time gravita-

77

tional red-shift was considered to be an alternative cause of the quasar red-shifts. However, there is nothing structurally peculiar about the quasars to indicate the presence of very strong gravitational fields. Indeed, a fair assessment of the situation is that no conventional explanation of the red-shift seems to fit the bewildering new data.

I now come to what may be potentially the strongest evidence against the steady state model. In 1965, as Dr Sciama has graphically described, Penzias and Wilson of the Bell Telephone Laboratories detected isotropic radiation in the microwave region. Further measurements by others have confirmed the existence of this radiation at many other wavelengths in the microwave, and the existence of this radiation is usually taken as evidence for the hot big bang.

The main reason for this is as follows. In the hot big bang model, of the type imagined by Gamow in his problem of nucleosynthesis, there is primeval radiation of very high temperature. Subsequently this gets diluted by the expansion of the universe and we should now expect to see a remnant radiation of low temperature. The intensity of this radiation over different wavelengths should follow the well-known black-body distribution. The observed radiation does seem to follow this law at long wavelengths. If we extrapolate it to short wavelengths it corresponds to a temperature of approximately 3° Kelvin (−270° Centigrade).

How can the steady state theory defend itself against this observation? First, it is necessary to point out that observations have not so far established the black-body character of the radiation. The critical range of wavelengths over which the observations are necessary to settle this point is not accessible to ground-based instruments. One has to rely on rocket flights. The rocket flights first seemed to indicate radiation far in excess of that predicted by the black-body law. Subsequent observations, however, do not seem to be inconsistent with the black-body law.

If the law is not one of black-body radiation, then its association with the hot big bang is not so compelling, and one can look for other explanations of its origin. A strong indication that this radiation may be due to other astrophysical processes operating in the universe comes from the fact that its energy density is comparable to that of cosmic rays, starlight, magnetic fields and some other energy sources which have no direct bearing on the big bang origin of the universe. Thus, one could argue that this background could arise from individual sources distributed all over the universe—just as the X-ray background is believed to arise from X-ray sources.

There is some difficulty with this explanation, however. The observed background is very homogeneous—it shows no patchiness at all. This means, the sources of such radiation must be very numerous and evenly distributed—far more so than ordinary galaxies. So far no obvious candidates have appeared for such sources.

The observed isotropy presents problems for the big bang models also—which is a fact not generally known. The question arises in this case 'Why should the radiation be so isotropic?' In the early stages of the big bang the region of communication is very limited, and it is difficult to see why regions far beyond the limit of communication of one another should have exactly the same amount of radiation. Attempts to answer this question in more complicated versions of big bang have not so far succeeded.

I have so far argued rather like a defence counsel trying to get his client acquitted for lack of sufficient evidence. I now come to other considerations which are less equivocal than the observational data I have discussed so far, and which clearly point to the superiority of the steady state theory over the big bang cosmologies. My own interest in the steady state cosmology has been sustained over the years, not because of the uncertain observational situation, but because of a remarkable physical property the theory has. I shall attempt to describe this property below, omitting the quantitative details.

One of the most puzzling facts in our everyday experience is the asymmetry exhibited by time. We are always conscious of the time axis being divided into two compartments—the past and the future, and we talk of events 'moving' from the past to the future but not vice versa. This consciousness has come out of observing events which cannot be reversed. For example, when we place a hot body in contact with a cold one, heat flows from the former to the latter, but not vice versa. When we set an alternating current circuit in operation, it radiates electromagnetic waves and loses energy—it never gains it. If such events are filmed and the film is run backwards in time we see new events which are never seen in real life. All these experiences are summed up by the statement that time has an arrow, from the past to the future. Why does time have an arrow?

To answer such a fundamental question we may be tempted to turn to basic physical laws. Perhaps there is something in these laws that produces the time arrow. But this is not the case. The laws themselves make no rule of this kind, they are time-symmetric. In other words, the events seen on the film running backwards in the above example, must be possible according to these laws. This makes the origin of a time arrow even more puzzling.

The usual way to get round this problem is to make an extra postulate which introduces this time asymmetry. The actual forms in which this postulate may be stated might differ somewhat according to different physicists—but the central idea is the same. Although this is a convenient device dealing with irreversible events it hardly takes us any further in understanding the origin of such events.

A more fruitful way is to look at such events in different branches of physics, and to see if there is any connection between them. The two examples described before occur in thermodynamics and electrodynamics, and are apparently not connected. I now give a third example which is even further apart from these two. This is the expansion of the universe. By saying that galaxies are moving further and further apart, we are introducing a time asymmetry into the picture. The reversed event, in which galaxies come closer and closer is not observed except in very isolated cases. Thus we have an arrow of time in cosmology, in addition to those in electrodynamics and thermodynamics described before. Are these arrows connected?

To a laboratory physicist, the idea that the arrows of time in electrodynamics and thermodynamics may be related, is not very unusual. But to him, to try to relate these two to the expansion of the universe, might seem fantastic. Yet, there exists a theoretical framework in which such a connection can be established.

To understand this connection, let us take a specific example—that of the alternating current circuit. Such a circuit radiates waves outwards. A simplified situation is shown in Fig. 11(a), where the source of the waves is at point A and the waves are spheres with A as centre. If we take $t = 0$ to indicate the time when the wave started outwards

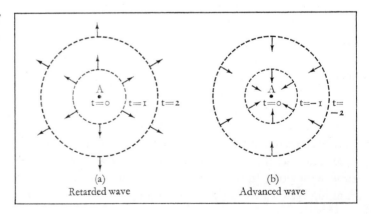

Fig. 11: Retarded and advanced waves.

(a) Retarded wave

(b) Advanced wave

from A. After one second (t = 1) it will be in the form of a sphere of radius one light-second. At t = 2 it has radius two light-seconds and so on. A wave of this type is called the retarded wave, since it reaches a point away from the charge at a *later* instant.

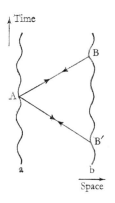

The time reversed version of the retarded wave of Fig. 11(a) is clear in Fig. 11(b). Here the wave goes out from A at t = 0, but reaches a point away from A at an *earlier* instant. Alternatively we may argue that the wave is converging on A. Such a wave is called the advanced wave.

Now, Maxwell's equations, which give us the fundamental laws of electricity and magnetism, tell us that advanced as well as retarded waves are theoretically possible. Nature, however, seems to reject the advanced waves altogether. Why? It is not possible to answer this question without introducing some extra postulate.

Fig. 12: Action at a distance.

There is, however, another way of describing electromagnetic theory which does not require an extra postulate but which brings in cosmology. This picture, known as *the action at a distance*, is in fact older than Maxwell's theory. In Maxwell's theory, two electric charges influence each other via the medium of a field. When one charge is set in motion it disturbs the field around it. This disturbance travels outwards (via retarded waves) and eventually reaches the other charge which then responds to it. In the action at a distance picture the two charges interact directly. In Fig. 12 there are two charges a and b moving in space and time. The influence of charge a from point A travels with the speed of light and reaches charge b at point B. The charge b reacts back in equal and opposite sense (according to Newton's third law of motion) along BA. Thus a retarded influence of a invokes an advanced response from B, and vice versa. The theory of action at a distance therefore treats advanced and retarded influences on an equal footing. In the figure, A is in interaction with two points on the world line of b, the point B is in the future and B' in the past relative to A.

The presence of signals going backward and forward in time is contrary to our experience and this has been one reason why this theory was never considered a serious rival of Maxwell's theory. It was not until 1945 that it was shown that this action at a distance theory can in fact be made to work. This remarkable work by two American physicists John Wheeler and Richard Feynman brought cosmology into the picture for the first time.

Wheeler and Feynman argued as follows. When we set a laboratory system A in motion, it interacts with the *entire universe* (see Chapter 8). Suppose we take the retarded wave

F

moving outwards from A at time t = o. It reaches a typical electrical charge B. As seen in the figure, this charge reacts via advanced waves which travel backward in time and arrive back at A at time t = o. Notice that the reaction from B arrives at A instantaneously after the emission of the wave by A, no matter what the distance of B from A! Therefore all the electric charges in the universe can, and do, react to the motion of A. Wheeler and Feynman then proceeded to calculate the reaction of the universe as a whole to the motion of a single electric charge in a laboratory. The results of their calculation can be described symbolically as follows.

When a single charge is set in motion, it emits signals equally via advanced and retarded waves. We can describe this by saying that the net disturbance from the charge is given by

$$F = \tfrac{1}{2} \text{(Retarded)} + \tfrac{1}{2} \text{(Advanced)}$$

The reaction of the universe, as calculated by Wheeler and Feynman, was

$$R = \tfrac{1}{2} \text{(Retarded)} - \tfrac{1}{2} \text{(Advanced)}.$$

The total disturbance leaving A is therefore

$$T = F + R = \text{(Retarded)}.$$

Thus we get only the retarded waves moving outwards, as observed. But, we also see that the retarded character arises because of the reaction R sent by the universe. This wipes out the awkward advanced part of F and augments the retarded part to the full observed value.

I have not so far said what model of the universe was used in this calculation. Wheeler and Feynman used the static Euclidean universe with a uniform distribution of electric charges. This model does not have a cosmological arrow of time, and this resulted in an ambiguity in the above result. Since the universe is time symmetric, we can reverse the direction of time in the above calculation and arrive at T = (Advanced)! This case was eliminated by Wheeler and Feynman by using thermodynamic arguments. I do not propose to go into the details, except to point out that this line of reasoning results in linking the arrows of time in thermodynamics and electrodynamics, with cosmology playing a secondary role.

However, a more realistic calculation should involve expanding cosmological models. Calculations in such models have been performed by various people. Although details differ slightly, the broad conclusions are similar, and startling. In ever-expanding big bang models we end up with T = (Advanced), i.e., opposite to what is observed! In the steady state model we get the correct answer T = (Retarded).

Big bang models which eventually contract, yield ambiguous answers.

The reason for this result is not difficult to see. To get the correct answer, we need the reaction

$$R = \tfrac{1}{2} \text{ (Retarded)} - \tfrac{1}{2} \text{ (Advanced)}$$

from the future half of the universe. In big bang models which are ever expanding, the matter density diminishes to zero in the future and there is not enough matter to produce the required R. In the steady state model, continuous creation of matter ensures sufficient matter density to produce the correct R. In big bang models, the solution $T = \text{(Advanced)}$ arises because there is enough matter in the highly dense past half of the universe to produce the opposite reaction

$$R = \tfrac{1}{2} \text{ (Advanced)} - \tfrac{1}{2} \text{ (Retarded)}.$$

In the steady state model, on the other hand, the density is not allowed to be large at any stage in the past, and this solution is avoided. In general, models of continuous creation of matter (which need not be in strict steady state) fare better than those with a big bang creation.

If we adopt this reasoning we have a powerful test of cosmological theories. It has the merit of being clear cut and free from the type of observational uncertainties we have been discussing in the middle part of this chapter. Further, we are also able to see a little more clearly how the arrows of time in different branches of physics are related. The ideas here have recently been extended to quantum physics with fruitful results. A discussion of these, however, lies outside the scope of this volume.

To summarise, the steady state theory has been under attack ever since it was formulated in 1948. Some of the criticisms levelled against the theory have subsequently been withdrawn as being incorrect, like the Stebbins-Whitford effect. Others have inspired important theoretical developments in astrophysics like the theory of nucleo-synthesis in stars. The theory has had its ups and downs, the worst period from its point of view being around mid-1965. At that time the observations of log N—log S and the Hubble diagram for early quasars looked to be against the theory and the discovery of microwave background had just been made. However, its fortunes have picked up since then as some astronomers have begun to appreciate the uncertainties involved in the quasar observations and the fact that we still do not know the nature of quasar red-shifts. The microwave background remains a potential threat to the theory and it is vital to know whether it has the character of a black-body distribution. *Until the observational situation clarifies further it is premature to write off the steady state theory.* Certainly, in the attempts to account for such fundamental concepts as the arrow of time, the steady state theory succeeds admirably whereas the big bang cosmologies fail miserably.

Cosmology thrives on controversies and these debates will and should continue. My own belief is that the structure of the universe is much more interesting and sophisticated than that suggested either by the big bang model or by the steady state model.

Despite the optimism of the preceding chapters, there are a great many things that the cosmologist not only does not know, but finds severe difficulty in envisaging a path towards finding out. Even if we beg the question of how the universe started, how did it become as it is now? In particular, how did the galaxies form? The encyclopaedias and popular astronomical books are full of plausible tales of condensations from vortices, turbulent gas clouds and the like, but the sad truth is that we do not know how the galaxies came into being. It is in such problems as this that cosmology is a paradigm of science tested to the limits, and our attempts to grapple with them have given rise to implications that concern science as a whole.

Many cosmologists are only too happy to leave the philosophical considerations of their work alone. One who is not is the Emeritus Professor of Astronomy, Sussex University, William H. McCrea.
L. H. J.

Professor
W. H. McCrea
FRS

Bondi defines cosmology as 'the field of thought that deals with the structure and history of the universe as a whole'. If we want to quibble about 'the universe *as a whole*', we could simply say 'that which we call the universe'. He justifies our regarding cosmology as a branch of the science of astronomy. As he remarks, there are various schools of thought on the subject of what is meant by a science. So far as I know, however, all scientists and philosophers of science agree that in a science what we are doing is to construct a theoretical or mental model that we compare with certain categories of experience of the actual world. We seek to establish a correspondence between a set of mental ideas related together by a system of logic, on the one hand, and a set of physical experiences on the other. We then go on to try to predict the results of experiments or observations, and we judge the success of our science by the accuracy and scope of its predictions.

This may sound all very straightforward, but anyone who ponders on such matters must realise that we cannot even begin to talk about them without a good deal of question-begging. To begin with, there is the fundamental problem of why the scientific procedure I have outlined should be possible at all! Why is it, for instance, that an astronomer who studies the sky tonight can go through certain operations on a piece of paper, or in a computing machine, and then say with the utmost precision what some other astronomer will see if he studies the sky say a thousand years hence? If you want an answer I will give you one. We have no idea! Without going into these deeper questions, however, what we learn from Bondi is that, in the general procedure of making a theoretical model and matching it up with what can be observed, cosmology is just like any other science.

While agreeing with Bondi about the procedures of science, I feel bound to take issue with him about his assertion that 'we do not honour scientists for being right'. Although it may not be quite according to Popper, I hold that science is gradually building up a body of what may be significantly called *understanding* of the physical world. The aim of a scientist, surely, is to contribute to this understanding. Unless he believes it is possible for him to do this, he would not want to dedicate his talents to the pursuit of science. Although the scientist knows that whatever he does someone else will ultimately do better, his hope is that his work will help this to come about. In this context there is a well-recognised sense in which a piece of work in science is either 'right' or 'wrong'. We may not dishonour a scientist simply because his work turns out to be wrong, but we do not honour him unless his whole endeavour is to be right.

It is precisely on these grounds that cosmology today lays itself wide open to attack. Having read the chapters in this book which deal with the study of the whole universe the reader is entitled to ask to what extent is there a picture emerging which furthers our understanding? And in a domain in which two apparently totally opposing theories can co-exist, with neither being able to invalidate the other decisively, what are we to say about the rights and wrongs of the situation?

The initial difficulty stems from the fact that the opposing theories arise from such eminently plausible concepts. In thinking about the universe in the large, we may begin by making either of two simple conjectures. We may suppose that there is nothing special about the epoch at which we see the universe and that, were we to see it at any other epoch, we should expect to see overall the same behaviour as we do now, even though we might be looking at some different material. Or we may think it simpler to suppose that, looking at the universe at any other epoch, we should see the same material as now even though it might show some different behaviour. Developing these two suppositions in the simplest manner possible, we obtain as alternative hypotheses about the working of the universe the steady state model and the big bang model.

As regards the creation of matter in these models in their simplest forms, creation in the steady state model has to proceed at the same mean rate everywhere and at all times, and every elementary particle results from its own spontane-

ous event of creation. In the big bang, creation occurs once
and for all in one unique spontaneous event. Both models
assume complete homogeneity in space; the steady state
assumes complete homogeneity also in time, while the big
bang assumes the greatest possible inhomogeneity in time.

There are two astonishing aspects of the record so well
presented by Sciama and Narlikar. On the one hand, there
are a number of qualitative features that these simple
models have in common and these agree well with the ac-
cumulating observations; no new observational result has
required the rejection of any such feature. The surprising
thing is that models that were first thought of as the simplest
possible for the exploration of certain ideas can now be
taken as serious descriptions of the large-scale behaviour of
the actual universe. On the other hand, the steady state and
big bang models are in other respects as far apart as any
could be. Here the surprising thing is that it took not far
off twenty years for the majority of cosmologists to become
convinced that observations almost certainly rule out the
steady state model.

The commonsense interpretation of this situation is as
follows. First, the general approach to the study of the
universe is sound and sensible. Were it on entirely the wrong
track, by now some observation would surely have produced
a blatant contradiction. Secondly, since observation has
found it so difficult to discriminate between the steady state
and big bang models, the difference between them may be
less drastic than it seems, and we might get a better model
by not going to such extremes. For instance, it might be
useful to consider models that have some non-uniformity in
both time and space. I shall, however, criticise this 'commonsense' view later on.

Meanwhile, let us consider a simple homogeneous model
universe suggested by the work of Wagoner, Fowler and
Hoyle and other related work. On the working hypothesis
that the actual universe behaves like the model, we infer the
following. The universe started in a hot big bang. The age
of the universe, i.e. the time since our part of the universe was
involved in the big bang, is approximately the Hubble time
given by the rate of expansion of our part of the universe,
which is between 10,000 million (10^{10}) and 20,000 million
(2×10^{10}) years. The material that survives the big bang to
provide the raw material for the subsequent formation of
galaxies is composed almost entirely of hydrogen and
helium. There was an early phase during which the universe
was radiation-dominated, before it became matter-dominated
as now. Shortly after the change-over, the matter became

highly transparent to the radiation. The background radiation that we now observe has been travelling almost unimpeded through the universe since that epoch. After the 'de-coupling' of matter and radiation, the temperature of the matter would fall more steeply than that of the radiation. There must have been a stage, roughly about 100 million (10^8) years after the big bang, when the mean temperature of the matter was only about 1 degree above absolute zero, and the mean density was about 10,000 times the present mean density through a large tract of the universe. This stage would be very favourable for the formation of condensations of matter by gravitational contraction—the most favourable in the history of the universe. While the model does not tell us that condensations must form nor, if they do, that they must be of galactic dimensions, it does thus provide a phase that is highly suitable for the formation of galaxies. Moreover, once galaxies have been formed as such, their radiation (not the background radiation) apparently would heat up any remaining diffuse material, which would also become even more diffuse as the expansion proceeds. Therefore the conditions in this material would have quickly become highly unsuitable for the formation of any further galaxies. Thus it is inferred that the formation of galaxies was restricted to a relatively brief phase in the history of the universe. A condensation of the raw material that is destined to form a galaxy like (say) our Milky Way, in its early life presumably experienced some of the exciting stages mentioned by Lynden-Bell in Chapter 4. For the reasons he states, effectively all the chemical elements, other than hydrogen and helium, that now form 1 or 2 per cent of the mass of such a galaxy must have been synthesised out of hydrogen and helium after the raw material had been formed into stars of some sort.

Judged in some ways, all this is a considerable amount of knowledge about the general history of the universe. It may be claimed to be compatible with much else in astronomy (for example the ages of stars inferred from ordinary astrophysics) and to be contradicted by no empirical result. Astronomers of fifty years ago would certainly have been profoundly impressed by this achievement.

On the other hand it is possible to adopt a quite different standpoint and to argue that the outcome is pitifully meagre. Astronomers can say that they have learned little more from cosmology than that the universe is a place in which we—our stars and our galaxies—could have been formed and in which we can go on existing. We might have been a little surprised had it been otherwise!

Even if this is the situation in the present state of the subject, we still have to ask whether we ought to expect a more adequate cosmology ultimately to explain much more about the astronomical universe than it has succeeded in doing up to now. In order to deal with this central problem, we have to consider the general properties of the objects that so persistently earn mention in this book—the galaxies.

The Milky Way galaxy is an assemblage of about a 100,000 million stars mostly occupying a lens-shaped region of space about a 100,000 light years in diameter. These stars shine with a total luminosity of about 10,000 million suns. The system contains also a rather patchy distribution of tenuous gas and dust making up some 2 per cent of its mass. Effectively all the radiation from the system comes from its member stars—our sun being a fairly average representative.

Throughout the observable universe there are other galaxies—probably some 1,000 million (10^9) to 10,000 million (10^{10}) can be seen with existing telescopes. Some giant galaxies may be ten times as massive as the Milky Way galaxy while some dwarf galaxies may be 10,000 times smaller. They all, nevertheless, bear sufficient family resemblance to each other to be treated as a single category. Quasars are about a thousand times less plentiful than galaxies but I shall treat them for present purposes as belonging to the same general category.

I can then make the following broad assertions: every scrap of matter that has ever been observed belongs to one galaxy or another. All radiation that has ever been observed —excepting only (in the opinion of most astronomers) the microwave background radiation and possibly the much feebler X-ray background radiation—comes to us directly from one galaxy or another. Nothing that happens to any galaxy is appreciably affected by anything else in the universe (unless we suppose, as Dr Sciama does in Chapter 8, that inertial properties are determined by the rest of the universe, but this would not affect the present argument). None of these assertions holds good for any lesser material systems—stars, for example, or any subdivisions of galaxies.

There may be a great deal of diffuse material spread out between the galaxies, but if so it could be detected only by some interaction with radiation from galaxies or, just possibly, with the background radiation.

In the early days of modern cosmology, it was fashionable to call the galaxies the 'building bricks' of the astronomical universe. It was, indeed, a convenient practical procedure to regard cosmology as the study of the universe of galaxies, in which each galaxy is treated as a single entity, and to regard the rest of astronomy as the study of what goes on inside each individual galaxy. When, subsequently, interest developed in clusters of galaxies, in the possible existence of intergalactic matter, and in the apparent actual existence of intergalactic (background) radiation, and so on, the unique status of the galaxies tended to take less attention. Nevertheless, as we shall see, this status is probably more significant than was appreciated at the outset.

This is because there cannot be astronomers to observe, nor objects for them to observe, until galaxies have evolved to about their 'present' state. And then the only things for an observer to observe are in his own or other galaxies! Once a galaxy has attained individual existence, however, its evolution is not appreciably affected by anything outside itself. Therefore, in general, the astronomer observes nothing that depends upon the properties of the universe in the large and none of his unsolved problems are likely to depend for its solution upon the properties of the universe in the large, i.e. upon cosmology. Conversely, of course, most astronomical studies will add nothing to our knowledge of cosmology. To put it bluntly, we learn from cosmology that we learn very little from cosmology!

All this is clearly in entire agreement with the rather disappointing situation I described earlier but at least the whole state of affairs is self-consistent!

That this is not an empty statement, in the sense that it could not have been otherwise, can be seen by recalling that when the expanding universe was first discussed it was thought that all kinds of things in ordinary astronomy might be accounted for by the fact that everything had been very much more compressed in the past. What we now realise is that almost nothing we can observe was directly affected by the past history of the universe in the large.

Basically this is why observation provides no direct contradiction of any generally plausible cosmological model and why, in particular, it took so long to reach a generally agreed choice between steady-state and big bang cosmology. Also it is significant that this choice was determined ultimately not by observations of anything to do with galaxies but by observations of what is held to be radiation surviving from a much earlier stage of a big bang universe.

A galaxy has a finite mass and so it contains a finite amount of energy; therefore it can have been radiating as we see it for only a finite time. Thus any body of material that now constitutes what we recognise as a galaxy must have been formed as such from material in some other state at a finite time in the past. Inevitably, therefore, we encounter the problem of the formation of galaxies.

If we list known unsolved problems of astronomy, with only one exception that I know of, we expect cosmology to be of no significance for their solution. The exception is the problem of galaxy formation. This is most satisfyingly in keeping with the outlook I adopt because, so far as this is concerned, there is cosmology but almost no astronomy before there are galaxies; after there are galaxies, there is astronomy but almost no cosmology. So, appropriately, we see the process of galaxy formation as the link between cosmology and astronomy. And, in fact, so long as we keep to the typical big bang model we think that we know when the galaxies were formed.

We might also expect cosmology to tell us why and how the galaxies were formed, why they have certain masses, spin momenta and linear sizes, and why they have whatever is their initial chemical composition, but this problem has not yet been solved. Moreover, those who have explored it most fully seem to be the ones who are most convinced that almost no progress has been made. I am not for one moment suggesting that the investigators have been wasting their time, because it is at least instructive to consider the principles on which the attempts have been made.

The most important reason why the problem of galaxy-formation has not been solved is that it has never been stated. Obviously we can state a problem only if we know where to start, and in this problem no one has yet discovered what should be used as starting conditions.

Traditionally, cosmologists have tried to start with a model universe that is assumed to be homogeneous in the sense of big bang cosmology and have asked how inhomogeneities can arise. The interest would be in inhomogeneities that would lead to condensations of material of galactic dimensions. What is immediately contemplated is that the homogeneous distribution of matter must be unstable with regard to some sort of disturbance, so that, a disturbance of sufficiently large extent will tend to increase in amplitude rather than to die away. Here we are dealing, however, with the universe as a self-contained system; so a disturbance in one region can be produced only by a disturbance in some other region of the system itself. This is ruled out if we suppose the whole system at any stage to be perfectly homogeneous in the accepted sense of a rigorous cosmological principle.

It is natural to say then that perfect homogeneity is the state we should least expect in nature. We certainly see tremendous irregularities in detail in the distribution of matter in the actual universe. Why then should we imagine that it ever was otherwise? Were the raw materials of the galaxies, just before they were formed, more diffuse than the materials of the existing galaxies (as most, but not all, cosmologists suppose), then the irregularities would indeed have been less extreme in the past than they are now. But it does seem absurd to suppose that nature first of all produced perfect homogeneity for apparently no other reason than to make it as difficult as possible to produce the observed non-homogeneity! Yet so soon as we admit the possibility of congenital inhomogeneities in the universe, we beg the question regarding the origin of inhomogeneities. We thus find ourselves in an unresolved dilemma as regards starting conditions.

Certain other general considerations also need to be mentioned. In many physical problems a characteristic mass or length may be isolated on purely dimensional grounds, but no such quantities are known in the case of the galaxies that would tell us that on any acceptable theory a galaxy must be a body of a certain mass and size. In consequence of this, and for other reasons, we expect the formation of galaxies to depend upon all the complicated atomic properties of real matter under a variety of physical conditions, not merely upon simple gross properties like density and pressure. Earlier chapters have reminded us that, when we look into the distance, we look also into the past, and so it might appear that with a sufficiently powerful telescope we could look far enough back actually to see a galaxy being formed. But Bondi reminded us also that we look back into a sort of 'fog' in which we can, in principle, not expect to see things clearly. Moreover, as we have remarked, before there were galaxies of some sort there were in any case no radiating systems that could be observed. It seems, therefore, that we could not possibly check a theory of very early stages of galaxy-formation by observation. But a theory has no scientific meaning unless it can, in principle, be checked by observation. So this is a case where we have to beware that a theory does not attempt to explain too much! We should perhaps, in the first place, restrict ourselves as far as possible to the problem: What was the matter that now forms a galaxy doing directly before that galaxy existed as such? Looking at the subject in this way, we need not then become concerned for the time being with the problem of the formation of clusters of galaxies.

Now some writers have discussed the possibility that some irregularity of density was present in the universe from the outset and that this led ultimately to the occurrence of galaxies. This idea has not achieved any success, since it assumes practically all that is to be inferred.

On the other hand, it is of the essence of any quantised system in physics for it to exhibit spontaneous fluctuations, and these may be amplified in sufficiently dense matter. An atomic bomb is almost an illustration of such amplification. In the early stages of a big bang universe the matter has greater density than ever again. It is natural to speculate, therefore, as to whether quantum fluctuations at such a stage could imprint upon the matter inhomogeneities involving amounts of the order destined to form galaxies. This would be an appeal to a species of fluctuation that we believe must occur, and is therefore quite different in principle from any arbitrarily assumed fluctuation. Indeed it is an alluring thought that all the greatest irregularities in the physical universe should be traceable to the unavoidable irregularities that occur in physics on the smallest possible scale. Unfortunately, such ideas are still only speculative and they have not yet produced a properly developed theory.

Ever since the modern idea of a galaxy became current there have been speculations about the possibility that the nucleus of a galaxy is a place where matter in its familiar state is produced from matter in some unfamiliar state—that the nucleus is a 'white hole' in recent terminology. Then this matter supposedly produces the galaxy concerned in its familiar guise. This has not led to any organised theory either, but the fact that such speculations have been seriously entertained shows how inadequate all other approaches have proved!

Bondi called attention to the striking fact that any region of the universe is obviously far from a state of thermodynamic equilibrium even though it has been in existence for such a long time. This is literally true. But what we normally call a state of thermodynamic equilibrium is the most probable state of the system concerned, subject to given applied conditions. For actual matter in an expanding universe it *might* be possible to prove that the state of resolution into galaxies is actually the most probable state, or at any rate that it is more probable than any other not too different state. Were this possible, then we should infer that the universe must be in this state, even if for the moment we cannot say how it got there. This would obviously be a big advance. The Argentine astronomer J. L. Sérsic does appear to have made some progress along these lines.

It is helpful to compare the problem of galaxy-formation with that of star-formation. Stars are still being formed in our galaxy, and in principle the process may be fully observed. Every astrophysicist is confident that it is only a matter of time until a satisfactory theory is developed, although one is not yet available. In principle the initial conditions on the raw material can be known as precisely as required, and the required laws of physics are all well known too. So in character it is just like any other problem in ordinary physics. Also, even if the process of formation is not yet known in detail, it is well understood why a normal star has mass, size, spin momentum, etc, within certain ranges of values.

We can then ask how a binary star, a star cluster, a galactic spiral arm, etc, is formed. In character, these are all ordinary physical problems like that of star-formation (though it may be significant that as we progress to larger systems, the relative ranges of values within which their properties must lie, become progressively less restrictive).

When we come to the problem of the formation of an entire galaxy the character has changed. In proceeding to larger systems, this is the first problem where this happens, and that is why I have discussed it at some length.

The difference is that cosmology has now become relevant and the immediate effect is that even in principle we can know less about initial conditions. As we have seen we do not yet know how to tackle the problem. Nevertheless I suppose every astronomer is convinced that some sort of solution will ultimately be found. But it must be a solution that relies less upon a knowledge of initial conditions than does the solution of a problem of ordinary physics.

We expect that as we go to still larger systems, clusters of galaxies and so on, this feature will become more and more significant. And this expectation seems to be clinched by the fact that if we want to discuss the formation of the universe, then obviously we have to do so without any initial conditions at all.

The classical and generally accepted view of science is that it explains something in terms of something else. The scientist makes certain postulates and draws certain inferences. If the inferences agree sufficiently well with certain experiences of the external world, he claims that these experiences are explained by whatever is assumed in the postulates. This is all he means by 'explained'. In fact, the development of science then consists in finding more experiences that are explained in this sense, or in finding simpler postulates from which the same results may be inferred, or both. Progress reduces essentially to explaining one set of postulates in terms of another set. In some recognisable way we require the latter set to be simpler or more fundamental than the former. And so the operation proceeds; step by step, the postulates are pushed further back. Each step, however, is in kind exactly like its predecessor. There can be no end to this process.

Nearly every scientist would, probably, assent to some such statement as a rationalised account of his activity. Nevertheless, if he analyses his motives he will almost certainly find some hope of finality, some instinct that, however remote, there is a goal before him. Otherwise, science is just an unending game played for ever according to unchanging rules. It is hard to believe that anyone would devote his life to what he himself is convinced is a game and nothing more.

Explicitly or implicitly, scientists have in fact contemplated other possibilities. One is simply the possibility that the described process of pushing the postulates further back somewhere comes to an end at a recognisable First Cause. Another possibility is that, as we go further and further back, the physical content of the postulates becomes less and less so that ultimately the whole of physics becomes a set of mathematical theorems. Such a possibility has appealed to some minds, but it is very hard to say what it means on further critical analysis.

There is also the possibility that as we go back the postulates become less and less important. This could happen in various ways. It might be that the content of the postulates tends to nothing as we go back from one set to another. Or it might be that the same consequences result from wider and wider sets of postulates until ultimately we get the same answer whatever we assume to start out with.

Science has not yet reached this state but the discussion in this chapter appears to show that it is tending this way and, if this is correct, it shows that the tendency should become more and more evident as we seek to work back from the formation of galaxies to the formation of the universe.

In the first place, we have done enough to see that, unless something like this situation is admitted, then the problems of the formation of galaxies and larger systems could apparently never be solved.

In the second place, it is a self-consistent view. In an ordinary physical problem, the equations concerned admit a range of solutions and we pick out the one we need by applying the boundary or initial conditions. The present conjecture is that, as we go to larger and larger systems in the universe, the range of solutions corresponding to any range of initial conditions becomes more restricted until, for the universe in the large, there is only one solution irrespective of initial conditions.

If this point of view is valid, it would thus imply that the actual universe is the only universe to which the laws of physics can be self-consistently applied. This would dispose of a basic misgiving about physics and cosmology that has often been expressed. It has usually been supposed that physics admits any number of physically possible universes whereas the actual universe is in a very significant sense unique. It seems highly unsatisfactory that the all-important property of actual physical existence would not single out one amongst all the universes that physics says could exist. The present contention—or conjecture—is that correct physics must admit only one possible universe.

As we have stressed, physics has not yet reached this state. Nevertheless, the trend is discernible in ways additional to those already described here. For instance, whereas the Friedman-Lemaître cosmological models were first presented merely as the simplest that could be thought of in order to illustrate certain types of behaviour, very sophisticated developments in physics suggest that no significant model could be very different. Again, while physics does not yet say that the universe at some stage must have a particular helium content, it is found that the helium content is highly insensitive to certain of the assumed 'initial' conditions.

Such trends may be deemed all the more significant because they depend upon very advanced developments in physics, many of which have been discovered only long after the cosmological models concerned were first thought of. If they are significant they would have to depend upon virtually the whole of physics. And even then we should not be in a position to assert that physics shows that the universe must be what it is observed to be. There would have to be interaction both ways and the utmost we could hope for would be to find that only one body of physical theory and only one model of the physical universe are compatible. In this sense, and in this sense only, we should reach the state of affairs which was expressed by the late B. C. Carter as 'Things are as they are because they couldn't be otherwise'.

The time is not many years past when students wishing to take up cosmology were deterred by their tutors because the subject, it was said, was 'useless'. The implication, often overt, was that since the distant universe had no measurable effect on us whatsoever, there was little point in studying it. This jibe was not only unfair, it was also untrue. Suppose you took a large pendulum that could swing in any direction, set it up at the North Pole and sat on the swinging bob for 24 hours, what would you observe? If you were a Bishop Berkeley or an Ernst Mach you would realise that your motion was guided by the constellations, while the earth rotated independently beneath your feet. Let Dr Dennis Sciama explain. L. H. J.

THE IN-
FLUENCE
OF THE
STARS
Dr Dennis
Sciama

On 13 February 1973 the scientific world celebrated the 500th anniversary of the birth of Copernicus. The controversy surrounding Copernicus's teaching that, despite appearances, the sun does not go round the earth once a year but that in reality it is the earth that goes round the sun; the parts played by Kepler and especially Galileo, culminating in Galileo's trial, now form one of the immortal chapters in the history of science. Newton, who was in fact born in the year that Galileo died, was puzzled by the proposal that *in reality* it is the earth that is moving round the sun. This proposal implies that there is a clear distinction between real motion and apparent motion. Newton did not find the distinction at all clear. To try to improve matters he emphasised two ideas. Firstly the real motion of a body has a *cause* in the form of a real force exerted by another body. Secondly, the motion so caused is an *acceleration* and not a velocity. These ideas are of course enshrined in Newton's laws of motion. At first sight they might seem empty (how can we tell that a force is acting except by observing an acceleration?). But they gain content when we add another idea not directly referred to in the usual statement of his laws, namely that the presence of a force exerted by a body can also be inferred from the existence and properties of that body.

We can now apply these ideas to the motion of the sun and earth. If we suppose that the earth goes round the sun we can easily find a cause for the earth constantly changing its direction of motion, namely the sun's gravity. This works out to be numerically reasonable. But if we suppose that the sun goes round the earth we get into trouble. The sun's motion round the earth can *not* be accounted for by the

Chapter 8

G

earth's gravity. If the earth had so much gravity, it would pull objects on to its surface much more strongly than it does.

Newton gave another argument for his point of view which left him open to criticism. This argument involved a related kind of motion to the revolution of the earth in a circle round the sun, namely the rotation of the earth about its own axis. This motion gives rise to similar problems. We know that the sun and the stars go round the earth once a day and again we want to say that this is only an apparent motion; in reality it is the earth that is spinning. Newton's laws of motion assure us that this is correct because they tell us that a spinning body should be flattened at the poles and bulging at the equator, and this the earth is observed to be. Another demonstration that it is the earth that is spinning is the motion of a Foucault pendulum, which is a pendulum hinged so that it is equally free to swing in any direction.

Imagine, for simplicity, the motion of a Foucault pendulum swinging at the North or South Pole. Because of the freedom of its suspension, the pendulum simply swings in a fixed plane, while the earth rotates underneath it. Viewed from the earth the pendulum appears to swing in a plane that slowly rotates, making a complete turn in a day.

Newton himself stressed another example of this idea with his experiment using a rotating bucket of water. For this experiment he hung a bucket of water from the ceiling by means of a long cord. By twisting this cord he was able to make first the bucket, and then the water, rotate. Then he held the bucket firm so that only the water rotated until it, too, was brought to rest by friction. From this Newton deduced that the water was rotating when it climbed up the sides of the bucket and the surface was no longer flat. This is an example of real rotation, or absolute rotation as Newton called it. The essence of the experiment can then be summed up in the statement that the shape of the water surface is independent of the motion of the water relative to the bucket, or, to generalise, absolute rotation has nothing to do with relative rotation.

This conclusion was strongly attacked by Bishop Berkeley. He claimed that Newton's experiment actually led to precisely the opposite conclusion. For why, asked Berkeley, did Newton lay such emphasis on the relative motion of the water and the *bucket*, since the role of the bucket in the experiment is simply to hold the water up and to provide a means for setting the water into rotation. There is plenty of other matter in the universe, and we have only to ask the question whether there exists any matter whose rotation relative to the water *is* required for the water surface to be curved to see immediately that the answer is indeed 'yes, the stars must be rotating relative to the water if the water surface is to be curved'. Berkeley concluded that it is the stars that are responsible for the physical effects that accompany rotation, and that there is no such thing as absolute motion.

In coming to this conclusion Berkeley showed himself to be ahead of his time. Nothing of any significance was added to the discussion until the advent of Ernst Mach, 150 years later. Mach's approach to the problem of absolute and relative motion was only a slight elaboration of Berkeley's, and it is important largely because it stimulated a rediscussion of the problem at a time when Newton's authority was unquestioned. In particular Mach stressed that absolute acceleration should always be replaced by acceleration relative to the stars, or as we would now say, relative to some suitably defined average of all the matter in the universe. On this view we can take the earth to be at rest, so long as we include amongst the forces acting on it those exerted by the stars and galaxies which, relative to the earth, have an accelerated motion. In particular the rotation of the material universe round the earth must be responsible for the flattening of the poles, the bulging at the equator and the motion of a Foucault pendulum.

This is a neat idea, but it immediately raises the question: how do the stars and galaxies do it? How do they exert this effect? This question was answered by Einstein in 1907. At that time he was much influenced by Mach's ideas and, rather unfairly to Berkeley, he referred to the replacement of absolute space by the material universe as Mach's principle. Mach himself did not give a detailed theory of how distant matter influences local dynamics. Einstein's answer was a suggestion of genius. He proposed that the effect was *gravitational* in origin. A rotating universe would have a different gravitational effect on the earth from a non-rotating universe, and it is this effect which shows up as the flattening of the poles and the bulging at the equator.

Why did Einstein make this proposal? In retrospect the answer is obvious, though it was not so at the time. The most striking property of the influence of the stars is that the motion it produces does not depend on the mass of the influenced object. For example, whatever the mass of a Foucault pendulum at one of the earth's poles, its plane of motion rotates at the same rate, namely once a day. It is clear then that the influence of the stars cannot be electric or magnetic in origin, because the accelerations induced by such forces do depend on the mass as well as the electric charge and magnetic properties of the body being acted on.

Indeed a body that is electrically and magnetically neutral is simply not acted on at all by these forces.

By contrast, there is no question of a body being gravitationally neutral. So far as we know gravitation acts on everything. Moreover, and this is more to the point, Galileo had discovered that, apart from air resistance, bodies of different mass have the same acceleration when falling in the earth's gravity. This was a well-known proposition by Newton's time. Its possible relevance to the influence of the stars was not suspected until Einstein proposed it in 1907. The point is clear; if a given gravitational force induces the same acceleration in all objects whatever their mass, and if an accelerating universe induces the same acceleration in all objects whatever their mass, then it is reasonable to assume that the influence of the universe is gravitational in origin.

But we can go further than that. We can understand *why* the forces concerned have this property. Consider for instance a body falling freely towards the earth. From the point of view of this body it is the earth and stars that are accelerating. Since the body is permanently at rest no net force can be acting on it; in other words the gravitational field of the earth must be cancelled by the gravitational field of the accelerating stars. But now the trick is done, because the equality of these two fields has nothing to do with the properties of the body. All that is required is that the stars have the right acceleration relative to the body to ensure that their gravitational field cancels the field of the earth. If we were to replace that body by another with a different mass the rest of the discussion would be exactly the same. The stars must have the same acceleration as before relative to this body, and so this body must have the same acceleration as the previous one relative to the stars. In other words, all bodies fall equally fast in the earth's gravity.

This consideration is not simply the physical basis of Mach's principle. It is also the physical basis of the general theory of relativity. To translate these ideas into complete mathematical form turned out to be a tricky business and it took Einstein from 1907 to 1915 to do it. Even then he was only partially successful in giving expression to Mach's principle. Let me give one example of his success in this respect. From what we have said about the effects of a rotating universe we would expect that when an individual body rotates, its gravitational field would drag around the plane of a Foucault pendulum on its surface. This effect is present in Einstein's theory. The angular velocity of this dragging is given approximately by the angular velocity of the rotating body multiplied by the gravitational potential it produces at the Foucault pendulum measured in units of the square of the velocity of light. For example, consider the neutron star at the centre of the pulsar in the Crab nebula. This star has a rotation period of 33 milliseconds. Now imagine that a Foucault pendulum were placed at one of the poles of the neutron star. Since this star is expected to have a mass similar to that of the sun but a radius of only about 10 kilometres, the gravitational potential at its surface is about one tenth, and the rotating neutron star would drag around the Foucault pendulum about three times per second. In other words if we sat on the neutron star and considered that a Foucault pendulum defined a plane fixed in absolute space, and then sat in such a frame of reference, we would actually see the whole universe going around about three times a second.

We must now ask the further question: leaving aside nearby neutron stars, does all the matter in the universe drag a Foucault pendulum on the earth around once a day or could this be done by something else—absolute space, for example? This is where the modern debates on the subject begin. The problem is technically difficult because Einstein's equations are non-linear. This means that the influence of many stars is not the simple sum of the influence of each one taken separately. It is, therefore, difficult to analyse the gravitational field of the universe in sufficient detail. However, it must be admitted that exact solutions of Einstein's equations are known which are not Machian and which do essentially involve absolute space.

There are three schools of thought about this situation. The first says: forget about Mach's principle. It may have been a useful guide for constructing general relativity, but now that we have the theory we can forget about its origins. Curiously enough Einstein himself took this attitude in later years. Only a week before his death I had the good fortune to talk to Einstein about this problem. I told him that I had come to defend his former self against his later self. He laughed, but remained unconvinced!

A second school of thought which would have left Einstein even more unconvinced, claims that his equations must be modified in order to give exact expression to Mach's principle. Well-known theories along this line are due to Brans and Dicke and to Hoyle and Narlikar.

The third school of thought is intermediate between these two. It points out that in solving Einstein's equations, which technically are differential equations, it is necessary to introduce so-called boundary conditions which essentially represent the influences acting on us that come, loosely speaking, from infinity. The theory as it is normally presented does not tell us how to choose these conditions, and the third school of thought believes that it is here that the Machian or non-Machian character of the solution is slipped in. In other words, the Mach principle should be a *selection* principle for choosing solutions of Einstein's equations that do not involve any element of absolute space. However, this is easier said than done because the non-linear character of Einstein's equations makes it difficult to separate out in a clear-cut and well-defined manner the influence of infinity from the influence of matter at a finite distance. Fortunately a recent technical development has made this possible, and thanks to work by a number of people including Alt'shuler, Lynden-Bell, Waylen, Gilman and Raine, a rigorous selection rule for Machian universes in general relativity is now available. In particular, Raine has shown that in uniform universes, that is, ones with no nearby rotating neutron stars and so on, the Machian selection rule does imply that the universe as a whole has zero rotation relative to a Foucault pendulum. So perhaps after all this chapter in the history of science has at last been completed.

But I would not want to end on such an abstract note. The reader is entitled to ask how well our final result compares with observation. With what precision can we say observationally that the universe as a whole is not rotating? Until a few years ago the best one could have said would have been that any rotation of the universe is so slow that its period must exceed about 10,000 million (10^{10}) years. Since for many purposes the universe has a time scale of about the same order this is not a very impressive result. However, as I point out in Chapter 5, a much more striking limit has been imposed since the discovery that the universe is filled with microwave radiation which probably has a blackbody spectrum at a temperature of 3° Kelvin, although the detailed spectrum is not important for the present argument. A rotation of the universe with a period of 10,000 million years would lead to a motion of the earth through this microwave background which would have been easily observed from the resulting Doppler effect. The upper limit on the earth's motion is observed to be 300 kilometres per second, and thanks to a beautiful analysis by Hawking, we know that this imposes a stronger limit than the previous one on the rotation of the universe by a factor of about 10,000. We can thus say with very high precision that the universe is observed not to rotate. And all good Machians would add that since there is no such thing as absolute space anyway the universe couldn't rotate even if it wanted to!

The heavy elements of which we are composed were forged in the nuclear furnaces inside giant stars. Yet many of the stars which now contain heavy elements are of the wrong type to have been able to produce them. The elements must have been ejected by a previous generation of supernovae and swept up by the stars of today. But computer studies show that to have produced the abundance of heavy elements we now observe, the mass of 'first generation' stars must have been considerably higher than now. What has happened to the missing stars? Some cosmologists are suggesting that up to 90 per cent of the matter originally present in the universe has disappeared forever down black holes.

In the following chapter, Roger Penrose, Professor of Mathematics, at University of Oxford, shows that these curious cosmological objects have a thoroughly respectable pedigree; one which can be traced in unbroken line to stars not very different from our own. L. H. J.

About 6,000 light years away, in the constellation of Cygnus the Swan, lies the blue supergiant star HDE 226868. Its mass exceeds that of our sun by a factor of about thirty, and its radius by a factor of nearly twenty-five. This in itself is nothing especially unusual. Many other stars of a similar nature are known. But once every five and one-half days, HDE 226868 circles in orbit about an invisible companion. It is this mysterious companion which concerns us here—with a mass one-half that of HDE 226868 but utterly tiny, its radius apparently being only about thirty miles! The companion of HDE 226868 is now believed by many astronomers to be a *black hole*—a bizarre consequence of the physical laws embodied in Einstein's theory of general relativity. Einstein's theory describes gravitation in terms of 'space-time curvature', and in a black hole, the gravitational field has become so strong that this curvature leads to some surprising and weird effects.

The identification of the companion of HDE 226868 as a black hole is not yet quite certain, but looks highly probable at the present time. There are other objects in the heavens which some astronomers claim are likely also to be black holes and it may be that the final definitive judgement on the existence of black holes will be centred instead on one of these. But however it comes, this discovery will be an event of the utmost importance to present-day physical theory. For theory predicts that black holes *should* exist and should occur sometimes as the end-point of stellar evolution. If black holes were found *not* to exist then this would point to some drastic revision necessary in the theory. On the other hand their *existence* will also pose fundamental problems

Chapter 9

for the theory and I shall attempt to elucidate some of these later in this chapter.

What is a black hole? For astronomical purposes it behaves as a small, highly condensed dark 'body'. But it is not really a material body in the ordinary sense. It possesses no ponderable surface. A black hole is a region of empty space (albeit a strangely distorted one) which acts as a centre of gravitational attraction. At *one* time a material body *was* there. But the body collapsed inwards under its own gravitational pull. The more the body concentrated itself towards the centre the stronger became its gravitational field and the less was the body able to stop itself from yet further collapse. At a certain stage a point of no return was reached, and the body passed within its 'absolute event horizon'. I shall say more of this later, but for our present purposes, it is the absolute event horizon which acts as the boundary surface of the black hole. This surface is not material. It is merely a demarcation line drawn in space separating an interior from an exterior region. The interior region—into which the body has fallen—is defined by the fact that no matter, light, or signal of any kind can escape from it, while the exterior region is where it is still possible for signals or material particles to escape to the outside world. The matter which collapsed to form the black hole has fallen deep inside to attain incredible densities, apparently even to be crushed out of existence by reaching what is known as a 'space-time singularity'—a place where physical laws, as presently understood, must cease to apply.

Since the black hole acts as a centre of attraction it can draw new material towards it—which once inside can never escape. The material thus swallowed contributes to the effective mass of the black hole. And as its mass increases the black hole grows in size, its linear dimensions being proportional to its mass. Its attractive power likewise increases, so the alarming picture presents itself of an ever-increasing celestial vacuum cleaner—a maelstrom in space which sweeps up all in its path. But things are not quite so bad as this. We are saved by the very minuteness of black holes—a fact which results from the smallness of the gravitational constant.

To see this, let us return to our picture of HDE 226868 (Fig. 13). Accepting the most recent figures for the dimensions involved, we have a black hole of some thirty miles in radius—in mutual orbit about a giant star whose radius is over 300,000 times larger. Despite its small size, the gravitational influence of the black hole is sufficient to distort the large star considerably out of spherical shape. It becomes rather like an egg whose small end is somewhat pointed in the direction of the black hole. A certain amount of material is dragged from this point and slowly falls inward to the black hole. It does not fall straight in, however. The black hole behaves much like a point mass. Most of the material dragged from the large star will remain circulating about the black hole for a long time. Only gradually, as frictional effects begin to play their part, will the material begin to spiral inwards. Again we must bear in mind the small size of the hole. (Imagine having to drain a normal-sized bath through a plughole a ten-thousandth of an inch across—or a bath the size of Loch Lomond through a normal-sized plughole!) The material can be only very slowly funnelled into the black hole. And as it gets funnelled in it gets compressed and very hot—so hot that the material must be expected to radiate light of very short wavelength, X-rays, in fact. Such X-rays are actually observed coming from the vicinity of HDE 226868. And the source of these X-rays (referred to as Cygnus X-1) appears, on the basis of detailed observations, to be in orbit about the visible component HDE 226868. The observed signals seem to be perfectly consistent with the black-hole picture I have presented. However we should remain cautious about drawing premature conclusions, as it is still conceivable that some alternative explanation of the observations may eventually turn out to be correct. The present evidence seems to be pointing ever more strongly in favour of Cygnus X-1 being a black hole, but even if for some reason this interpretation does turn out to be erroneous after all, it would still be very surprising

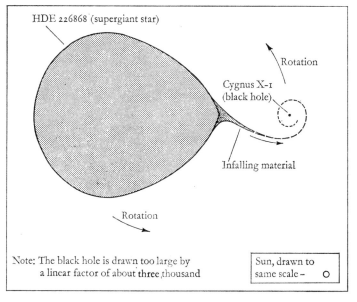

HDE 226868 (supergiant star)

Rotation

Cygnus X-1
(black hole)

Infalling material

Rotation

Note: The black hole is drawn too large by
a linear factor of about three thousand

Sun, drawn to
same scale – O

Fig. 13: The gravitational field of the black hole distorts the supergiant star out of spherical shape and drags material from it.

(on the basis of present theory) if *no* black holes were found to exist. To indicate why, I should explain something of the picture that astronomers and astrophysicists have developed concerning stellar evolution and then indicate some of the theory that lies in support of the black hole picture that I have presented.

Let us consider first what theory and observations tell us to expect for the future of our sun—or of any other normal star of about the same mass. After shining at approximately its present brightness for about 7,000 million years the sun will begin a change which will transform it beyond recognition. According to the well-accepted theory of stellar evolution the sun will grow to an enormous size and become, like stars such as Antares in the constellation Scorpio, a red giant some 200 million miles in diameter. By this time the planets Mercury, Venus and the earth will have been burned away and their former orbits will lie well within the new solar surface. The density of the sun's material will by then have fallen from its present value of a fifth that of the earth to a tenth of the density of air!

As it continues to burn more and more of its nuclear fuel, making heavy inroads on its helium and heavier elements as well as hydrogen, the bloated sun will halt its expansion and begin to contract—down past its present size, smaller and smaller until it stabilises as a white dwarf star perhaps about the size of the earth. At this stage further contraction will be impossible because the electrons of its atoms will be packed together so closely that a law of quantum mechanics known as Pauli's Exclusion Principle will come into play. This principle states that no two electrons in an atom can occupy the same energy state. One can envisage the atoms so closely squeezed together under the dwarf sun's tremendous gravitational field that, pictorially speaking, any closer spacing would force the electrons to get in each other's way so that they cannot, at these temperatures and pressures, be squashed together any further. In this state the density of the solar material will be such that a matchbox full of it would weigh several tons. No material on earth has a density remotely approaching that of a white dwarf but as with red giants, many white dwarfs can be seen in our galaxy. Their ultimate fate is simply to cool off to form black dwarfs and thereafter act merely as centres of strong gravitational attraction. The planets, Jupiter, Saturn, Uranus, Neptune, Pluto and possibly Mars, will still continue to circle the ancient sun aeons after it has died.

White dwarfs are part of the normal evolutionary history of average-sized stars like the sun, and astronomical observations show actual stars at each stage of stellar evolution through the stage now reached by our sun, on to the red giant phase and back to white dwarfs. Moreover the theory of stellar physics fits these observations closely, but not all stars can follow this 'normal' evolutionary path.

As long ago as 1931, Subrahmanyan Chandrasekhar calculated that there must be a maximum mass above which a white dwarf could not sustain itself against even further gravitational contraction despite Pauli's Exclusion Principle. Furthermore, this mass was not much greater than the mass of the sun—1.4 times according to Chandrasekhar; a little less according to more recent computations. Many stars have masses considerably more than 1.5 times that of the sun. What is going to happen to them?

The answer depends on just how heavy the star is. Consider a star of twice the mass of the sun. Like the sun it will also expand to an enormous size and then recontract, but being more massive than Chandrasekhar's limit for a white dwarf it will be unable to settle down to final equilibrium in the white dwarf state. To picture what happens it will be useful to consider the giant phase of a star more fully. As soon as the central density of the star reaches that of a white dwarf, the outer layers of the star expand, and they go on expanding as more and more of the central material gets compressed into a white dwarf state. So the giant star develops a growing white dwarf core. In the case of the sun, all the material that remains in the star will eventually become part of this white dwarf. But if the star is too massive, there comes a point at which the core effectively exceeds Chandrasekhar's limit, whereupon it promptly collapses. In the process of collapse there is a tremendous release of energy, much of which is in the form of neutrinos which are absorbed (so it is believed) in the outer regions of the star, heating the envelope to an enormous temperature. A cataclysmic explosion ensues—a supernova explosion— which blows off a considerable proportion of the mass of the star. Many supernovae have been seen, the last one in our galaxy being described by Johannes Kepler in 1604. A supernova will outshine a whole galaxy for a few days and may even be visible by daylight. As much as 90 per cent of the star's mass might be thrown off by the cataclysmic explosion to form tenuous debris of heavy elements which may later enrich the stars of a second generation. But especially interesting is the collapsed remnant of the star left behind at the centre of the rapidly expanding cloud of ejected gases. This core is much too compressed to form a white dwarf and it can only find equilibrium as a neutron star.

A neutron star is tiny even by comparison with a white dwarf. Indeed the reduction from a white dwarf to a neutron star is even more than the reduction of 100 to one from the sun to a white dwarf, or probably rather more than the approximate reduction of 250 to one from a red giant to the sun. A neutron star may be only 10 kilometres in radius or only about one seven-hundredth the radius of a white dwarf. The density of a neutron star could be more than a hundred million times the already extraordinary density of a white dwarf.

Our matchbox full of neutron star material would now weigh as much as an asteroid a mile or so in diameter. The star's density would be comparable with the density of the proton or neutron itself—in fact a neutron star could in some ways be regarded as an over-sized atomic nucleus, the only essential difference being that it is bound together by gravitation rather than by nuclear forces. Individual atoms have ceased to exist as such. The nuclei are touching and form one continuous mass. As for the electrons, as Professor Lynden-Bell has explained in an earlier chapter, what has happened is that the stupendous gravitational forces have squeezed the electrons into the only space available to them —that already occupied by the protons, reversing the usual reaction so that the star is now composed mainly of neutrons and it is the Pauli Exclusion Principle acting on these neutrons that supplies the effective forces preventing further collapse.

This picture of a neutron star was predicted theoretically by the Soviet physicist Lev Landau in 1932 and studied in detail by J. Robert Oppenheimer, Robert Serber and George M. Volkoff in 1938 and 1939. For years many astronomers doubted whether neutron stars could actually exist. Since 1967 the observational situation has changed dramatically, because in that year the first pulsars were observed. Since then the theory of pulsars has developed rapidly and it now seems virtually certain that the radio and optical impulses emitted by pulsars owe their energy and extraordinary regularity to the presence of a rotating neutron star. At least two pulsars reside inside supernova remnants, one being the Crab Nebula, and this gives further support to the theory that pulsars are in fact neutron stars.

There is a maximum mass above which a neutron star would not be able to sustain itself against still further gravitational contraction. There is some uncertainty as to the exact value of this maximum-mass limit. The original value given by Oppenheimer and Volkoff in 1939 was about 0.7 times the solar mass. More recently, larger values of up to three solar masses have been suggested. These higher values take into account the idea that the heavy subatomic particles called hyperons may be present in addition to ordinary neutrons and protons. But there are stars whose mass is more than fifty times the mass of the sun. What will happen to these? It seems exceedingly unlikely that all these stars would, as a result of their final collapse phase, or earlier, inevitably throw off so much of their material that their masses would always fall below the limits required for a stable white dwarf or neutron star to be the result. In these the neutron core would be unable to remain in equilibrium and would have to collapse further inwards. But what other forms of condensed matter might be possible considering that here we have densities in excess even of the fantastic value that is maintained inside a neutron star?

In this case, theory tells us a different story: although greater densities can be achieved, it is not possible to obtain any further stable final equilibrium configurations. Instead the gravitational effects become so overwhelming as to dominate everything else. Newtonian gravitation theory becomes quite inadequate to handle the problem, and instead, we must turn to Einstein's theory of general relativity. But in so doing, we are led to a picture so strange that even the phenomenon of the neutron star must seem commonplace by comparison. This new picture is the one which has now earned the description of a black hole.

Briefly, a black hole is a region of space into which a star (or collection of stars or any other bodies) has fallen, but from which no light, matter or signal of any kind can escape. Before examining this picture in some detail, consider (Fig. 14) the degree of further contraction that would be necessary for a neutron star to be compressed down to the

Fig. 14: The neutron star illustrated has the same mass as the sun but has a diameter 700 times smaller than the earth.

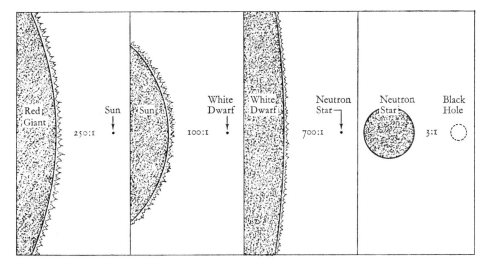

| Red Giant | 250:1 | Sun | Sun | 100:1 | White Dwarf | White Dwarf | 700:1 | Neutron Star | Neutron Star | 3:1 | Black Hole |

size of a black hole. We have already seen that from the sun's dimension down to that of a neutron star involves a contraction of about 70,000:1 in linear size; from a red giant to a neutron star, a linear contraction of about 20,000,000:1. In view of this, it is perhaps surprising that a further contraction of only about 3:1 is required for a neutron star of one solar mass to become a black hole—in this case, with a radius of about two miles. Larger black holes are also possible, the radius of the hole being proportional to the mass, so they would result from the collapse of a body more massive than the sun. For example, Cygnus X-1 appears to be about 15 times more massive than the sun, so its radius (assuming little rotation) is about 30 miles.

The reason I emphasise the slight nature of this further contraction, only 3:1, is that faced with the unsettling nature of the black holes, people have naturally asked whether our physical theories are tenable under these extreme conditions. But these theories seem to have worked well in describing a very large range of stars of enormously different sizes and densities. In any case, the conditions under which a black hole is formed are not so extreme as all that— not necessarily more extreme than the situation of a neutron star. For example, the densities involved as the collapsing star crosses the absolute event horizon are not vastly different from those inside a neutron star. The larger the collapsing mass, the less would be this density—less in inverse proportion to the square of the mass. In the case of Cygnus X-1 this density would be rather less than that of a neutron star. It has often been considered by astronomers that collections of mass of up to 100 million suns or more might be involved in gravitational collapse in galactic centres. The density at the time such a huge mass crosses the event horizon might then be only about that of water. So the local conditions need not be excessive when a black hole is formed and there seems no reason to suppose that the black hole condition might render general relativity somehow inapplicable.

On the other hand it must be admitted that general relativity plays anything but an irreplaceable role in observational astronomy. It is still possible that general relativity might be wrong; after all, the experimental tests of general relativity which have been successfully performed are still not very numerous. Although no conflict between the theory and observations exists at present (any apparent conflict being explicable by other means), the observations do not point *conclusively* in the direction of general relativity. There is still scope for alternative theories of gravitation.

But no one would deny that general relativity *is* an excellent theory; almost certainly the most satisfactory theory of gravity available to us. Furthermore the theory which at the present time may be regarded as general relativity's most serious rival (namely the Brans-Dicke-Jordan scalar-tensor theory), itself leads to a black hole picture nearly identical to that arising in Einstein's theory. Even in Newtonian theory a phenomenon similar to that of a black hole may be said to occur. As long ago as 1798, Pierre Simone Laplace used Newtonian theory to predict that a sufficiently massive and concentrated body should be *invisible* because the escape velocity at the surface could be greater than the speed of light. A photon or particle of light emitted from the surface of the body would simply fall back to the surface, it would not escape to be observed at large distances from the body. This description is perhaps arguable, but it shows that there is a situation to be faced, even in Newtonian theory.

I only make these observations to show that if one's sole motive for casting doubts on the tenability of general relativity is to get out of the black hole situation, one might just as well stick with relativity, because its rejection will not necessarily make the black hole go away. So I propose, from here on, to restrict my discussion to considerations solely within the bounds of Einstein's general theory of relativity.

To begin with, consider the standard picture of a non-rotating black hole. The black hole is characterised by a spherical surface whose radius is proportional to the hole's mass. This surface is called the 'absolute event horizon'. Its defining property is that signals emitted inside it cannot escape, whereas from any point outside it signals can be emitted that either do escape or fall into the black hole. The radius of the event horizon can be calculated by multiplying twice the mass by the universal gravitational constant and dividing the result by the speed of light squared. Performing this calculation for the sun yields the result that the sun would have to be collapsed into a sphere four miles in diameter if it were to form a black hole. The absolute event horizon would be the surface of this four-mile sphere.

The body whose collapse is responsible for a black hole's existence has fallen deep inside the event horizon. The gravitational field inside the event horizon has become so powerful that even light itself is inevitably dragged inward regardless of the direction in which it is emitted. Outside the event horizon light escapes if it is aimed suitably outward. The closer the emission point is to the event horizon, the more the wave front of the emitted signal is displaced back toward the centre of the black hole. We may intuitively regard this displacement as being caused by the effect of the gravitation on the motion of the light. Owing to the intense gravitational attraction of the black hole, light travels more easily in the direction of the black hole than in the outward direction. Inside the event horizon the inward pull is so strong that outward motion of light has become impossible. This applies not just to light but to any signal originating within the black hole. As for a photon emitted radially outward from the surface of the black hole, on the event horizon, it will mark time, forever hovering in the surface itself at the same distance from the centre of the black hole.

It may seem that this is odd physics indeed, quite unlike the normal situation in relativity theory, where the speed of light has always the same constant value in all directions. But, strange as it may seem, the local physics in the neighbourhood of the absolute event horizon *is* the same as elsewhere. An observer at the event horizon who tries to measure the speed of light must himself be crossing the horizon by falling inwards into the hole. To him the speed of the light hovering on the horizon is indeed the same constant value, in the outward direction.

It would be natural for a reader who is not familiar with general relativity theory to find such a situation confusing. This is partly because so far we have been using a purely spatial description rather than a space-time one—and for many purposes a space-time picture is more illuminating than a spatial one. Strictly speaking, a space-time picture needs to be drawn in four dimensions, but an overall description of the space-time situation can be obtained by suppressing one of the spatial co-ordinates in the space-time diagram and substituting a time co-ordinate. This gives an instantaneous picture of what is going on at all times, and obviates the need for many sequential 'snapshots' of a developing situation.

Consider a light flash emitted in all directions from a given point in ordinary space. The wave front of the flash would be a sphere centred on the emitting point and growing larger each moment at the speed of light. A purely spatial representation of the flash would be a sequence of spheres (Fig. 15) each sphere larger than the preceding one, marking the position of the light flash's spherical wave front at a given moment in time. A space-time representation of the light flash, however, would be a cone whose vertex represents the time and place at which the light flash is emitted, the cone itself describing the history of the light flash.

By the same token the history of a star's collapse down to a black hole can best be depicted in a space-time representation (Fig. 16). The locations of the light cones at various points in space-time show how light signals propagate in the gravitational field. At some points the light cones are drawn as being tipped over, but this is not something that would be noticed by a local observer. Such an observer would follow a path in space-time that proceeds into the interior of the light cone; his speed can never be greater than the locally measured speed of light, and only inside the light cone is this criterion met. But the tipping of the light cones does affect what an observer at large distances can see. Fig. 16 shows that material particles and light signals which originate inside the event horizon are inevitably driven further inwards. For a particle or signal to cross the event horizon from the inside to the outside it would have to violate the condition mentioned above; it would have to exceed the local light speed, in violation of relativity.

By taking a horizontal section through the space-time picture we get a spatial representation of the situation as in

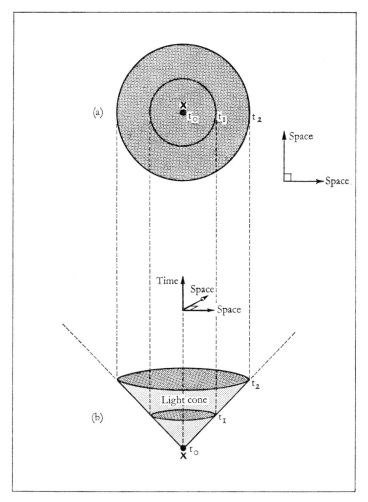

Fig. 15:
(*a*) *propagation of light from x in space at times t_1 and t_2.*
(*b*) *propagation of light in space-time to give a 'light cone'.*

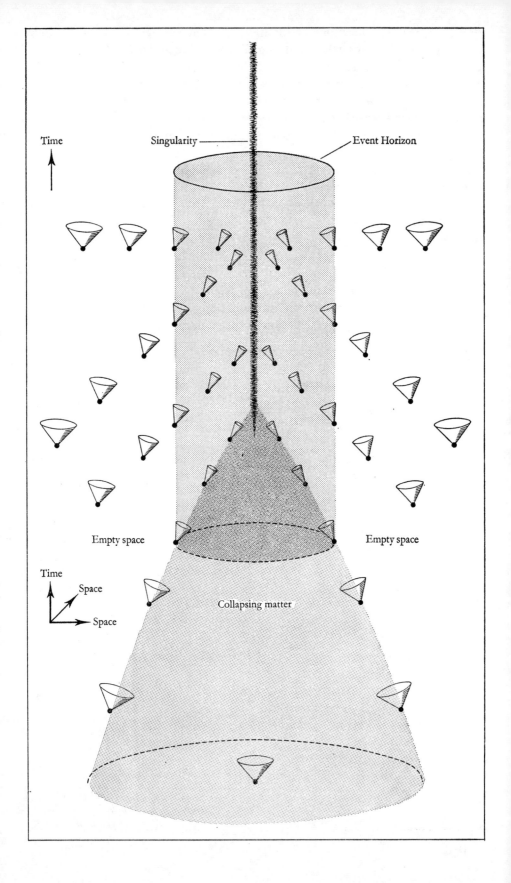

Fig. 17. Perhaps the main advantage of a spatial representation, apart from its greater familiarity, is that one need not dispense with the third spatial dimension when it is important not to do so. The light cones can be depicted as points (the origin of a light flash) surrounded by sphere-like surfaces (the location of the light flash a moment later)—except that when the light cone is tipped over, these surfaces do not actually surround the point of origin. In this later case it would be necessary to exceed the local light speed in order to 'stay in the same place'. A serious drawback with such spatial pictures, therefore, is that it becomes hard to interpret situations of this kind. If a space-time description is used, then it becomes easier to accept that the local physics is the same whether or not the light cones are depicted as tipped over, the 'tipping' being merely an aspect of the local description. (Imagine the figure drawn on a rubber sheet; then the light cone could be twisted around vertically without disturbing much of the rest of the picture.)

Although nothing can ever get out of a black hole, things can fall in. Indeed, it is quite possible that stellar astronauts traversing the depths of space in ages to come will run precisely this risk. Not that they will be likely to encounter a black hole by accident—the smallness of black holes compared with the vastness of the universe will see to that. Indeed, they would have to seek out a black hole deliberately if they wished to experience this 'ultimate trip', and this might be difficult because black holes are not directly visible.

And what will happen to a hapless astronaut who falls into a black hole? What, indeed is the fate of the original body which collapsed to produce the black hole? Assuming that the exact spherical symmetry is maintained right down

Fig. 16 (left): Space-time representation of collapse of a spherically symmetrical body to form a black hole, showing propagation of light inside and outside the event horizon.

Fig. 17: Light propagation inside and outside a non-rotating black hole.

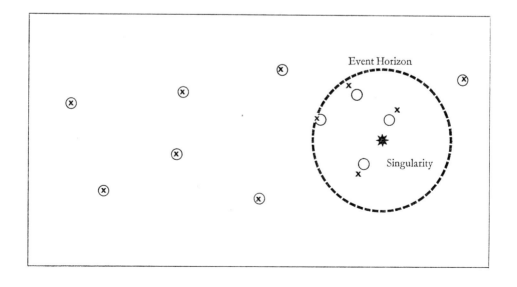

Event Horizon

Singularity

to the centre, the answer provided by general relativity is an alarming one. According to the theory, the curvature of space-time increases without limit as the centre is approached. Not only is the material of the original body squeezed to infinite density at the centre of the black hole—and crushed, effectively, out of existence—but also the vacuum in space-time which is left behind by the body, itself becomes infinitely curved. The effect of this infinite curvature on an observer, were he foolhardy enough to follow the body inwards, would be that he experiences mounting gravitational tidal forces—tidal forces that mount rapidly to infinity.

The gravitational tidal effect is the most direct physical manifestation of space-time curvature. Einstein pointed out that the gravitational force on a body can be eliminated at any one point simply by choosing a frame of reference that is falling freely. He gave the famous example of a lift that broke its cable and fell toward the earth. Any passenger inside would be falling at the same rate as the lift, so he would feel no net gravitational force relative to the lift, and, indeed, would float free of gravity inside it. Such elimination of the gravitational force by free fall is now a familiar feature of space travel. The tidal effect, however, cannot be so eliminated and is, therefore, an absolute manifestation of the gravitational field. Imagine an observer falling freely in the earth's field. Suppose he is surrounded by a sphere of particles that he initially observed to be at rest with respect to himself. The Newtonian gravitational field of the earth varies as the inverse square of the distance between it and any other body, pulling more strongly on objects closer to its surface than on objects farther away. This non-uniformity of the gravitational field tidally distorts the sphere of particles into an ellipsoid which bulges out in the direction of the earth and in the direction away from it. The earth's ocean tides are a familiar example of an effect of this kind, where in this case the earth is experiencing the tidal effects of the moon and the sun.

Fortunately for us, the tidal effects due to gravity encountered in the solar system are small. Nobody complains that his feet experience more gravitational pull towards the earth than his head. There is a difference, but it is not noticeable in the ordinary way. The space-time curvature responsible for this difference has a radius of about the distance from the earth to the sun. (This is a pure coincidence since the sun is itself irrelevant to this particular tidal effect.) At the surface of a white dwarf, on the other hand, the space-time curvature is considerably larger, the radius of curvature being of the same order as the radius of the sun.

This tidal effect would be very noticeable to an astronaut in orbit around the white dwarf. In fact, his head and feet could experience a difference in forces of perhaps one-fifth of the total force the astronaut normally experiences standing on the earth's surface. At the surface of a neutron star, however, the tidal effect is, by ordinary standards, enormous. The radius of space-time curvature there is only about thirty miles. It is clear that no astronaut in a low orbit around a neutron star could possibly survive. For even if he curled himself into a small ball, the gravitational acceleration at various parts of his body would differ by several million times the gravity at the earth's surface.

However, instruments could in principle be built to withstand such high tidal forces—all that would be necessary would be to make them so tiny that the difference between the gravitational attraction on the side nearest the neutron star and the side furthest away is small.

Suppose our astronaut carries such a tiny, rugged instrument as he flies towards a black hole of one solar mass. Long before he reaches the event horizon he will be destroyed by the tidal forces, but his instrument will survive intact as it crosses the event horizon, where it experiences tidal forces about thirty times those at the surface of the neutron star. But it survives only for a while, because the tidal forces encountered in a black hole rise to infinity at the centre. As the instrument falls in towards the centre, the mounting tidal forces will rise rapidly, ripping to pieces in turn the material of the instrument, the molecules of which this material is composed, the atoms which constitute these molecules, the atomic nuclei, and finally the fundamental particles which a moment ago had been the building-blocks of these nuclei. And the entire process would not last more than a few thousandths of a second!

So anything falling into a black hole, whether it be a space ship, a hydrogen molecule, an electron, radio-waves or a beam of light can never emerge again. So far as our universe is concerned it disappears completely and forever into nothing. But how can this be? Is it not a basic law of nature that matter or energy can never be completely destroyed but only converted from one form into another? The question is a perfectly respectable one, but it can be shown by rigorous argument, based on general relativity, that there must be a region inside a black hole, a region of infinite curvature, called a space-time singularity at which the known laws of physics break down. So there is no known conservation law that can be relied on at the centre of the black hole. There is no reason known why the matter should *not* just be totally destroyed as it reaches the singularity. Eventually, perhaps, laws of nature may be formulated which govern the behaviour of space-time singularities, but no such laws are known at present.

This situation has a more familiar manifestation in a somewhat different context. General relativity, like virtually all viable physical theories, is reversible in time. So corresponding to any solution of the equations in which time runs one way, there must be another in which the time-sense is reversed. This leads us to expect that the above situation could—in principle—exist in a time-reversed form. Initially there would be the space-time singularity. Then matter would appear: elementary particles, light. Only later would these particles collect together into atoms, molecules or stars. In fact, a picture of this kind has been considered for many years by astronomers and cosmologists as a model of the creation of the universe. The initial big bang of the cosmological models is, like the centre of a black hole, also a space-time singularity, where the curvature of space-time becomes infinite. But now, rather than being destroyed, matter is created at the singularity. The cosmological big bang is not precisely the time-reverse of a black hole, however, since the singularity is all-embracing, unlike the relatively localised singularity inside the black hole. The basic difference is one of size, and we may indeed envisage more localised 'little bangs', called *white holes*, which are more precisely the time-reverses of black holes. A number of theoreticians have considered such white holes seriously in connection, in particular, with models for quasars. However, I must say that I personally regard the possibility of the existence of white holes with considerable unease. The reason is basically this. Once a black hole is formed there is apparently no means of destroying it. It is created violently, but then settles down and sits around forever—or until the universe re-collapses at the end of time. Now a white hole—the time reverse of a black hole—would have had to have *been* there since the beginning of time—tamely and invisibly biding its time before making its presence known to us. Then, when its moment arrives, it explodes into ordinary matter. But this moment is of its 'own' choosing, governed, apparently, by no definite law. Of course, there is no fundamental reason why this should not occur; the idea simply seems untidy, at variance with thermodynamics and probably also with observation. Nevertheless we *are* still stuck with the big bang and that also seems untidy. But here there appears to be no way out.

But let us now return from the abyss of speculation and

pursue the argument concerning black holes. Quite apart from the doubts I have already raised about the validity of the general theory of relativity, there are other questions that need to be settled before one can fully accept the theoretical concept of the black hole as a realistic description of something that actually occurs in nature. In the first place, can we be sure that enough is known about the nature of matter under the extreme conditions required to form a black hole for the predictions to carry conviction? What role does the assumption of exact spherical symmetry play in the discussion? To what extent does the black-hole picture fit in with astronomical observations? Let us consider these questions in turn.

As I have already pointed out, the densities involved in the formation of a black hole need not be excessive. The same applies to space-time curvatures. A black hole with a mass ranging from 10,000 to 100 million solar masses is sometimes considered a possible candidate for what might inhabit the centre of a galaxy, and a collapsing mass equal to 100 million solar masses would reach a black hole situation when its average density was roughly the density of water. The tidal effects at the event horizon are, like the density, proportional to the inverse square of the mass of the hole, so for a body of 100 million solar masses these tidal effects would be somewhat less than those produced at the earth's surface. As an astronaut passed through the absolute event horizon he would notice nothing. He would have no means of telling that an irretrievable situation had developed, because the exact location of the horizon is not something that can be discerned by local measurement. After this he would have but a few minutes to enjoy the experience of life inside a black hole before the tidal effects mounted to infinity. In the case of a black hole of 10 thousand million solar masses he would have about a day.

The question concerning the role played by the assumption of spherical symmetry has to be examined more carefully. If we do not assume spherical symmetry, then we cannot appeal to the exact solutions of Einstein's equations on which we have based the foregoing discussion. Furthermore, even if we assume that initially the deviations from spherical symmetry are slight, we should have every reason to expect that near the central point these asymmetries would be enormously magnified. Might not the different portions of the collapsing body miss one another? Perhaps they could re-emerge after a close encounter and bounce out again. Even if they did not, can we say anything about the final configurations of the gravitational field resulting from the collapse? It is fortunate that, owing to some general theorems that have been proved over the past few years, a remarkably complete picture of asymmetrical collapse has emerged.

Considering the picture in a little detail, suppose that a massive star or a collection of bodies collapses and that deviations from the spherical symmetry are at first comparatively small. We can establish that a point of no return has been passed if a certain criterion is satisfied. That criterion can be stated in several different ways, but the following is the simplest. Imagine that a flash of light is emitted at some instant at some point in space. The flash of light will follow the light cone centred on the point according to our space-time representation (Fig. 18). The light rays start out from the point by diverging in all directions. When they pass through matter or through a gravitational field, the matter or the field has a focusing effect on the rays. If enough matter or a sufficiently strong gravitational field is encountered, the amount that the rays diverge can be reduced to such an extent that this divergence is actually reversed, that is, the rays start to converge. The required

Fig. 18: The birth of a black hole.

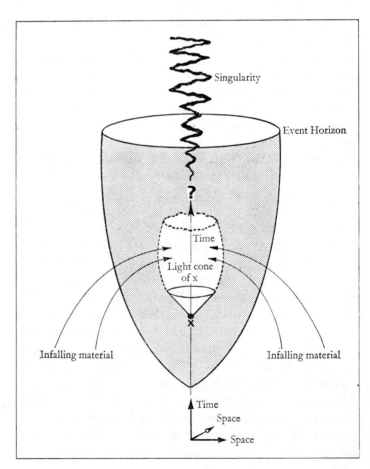

criterion for a point of no return is that every light ray from the space time point encounter enough matter or gravitation for the light cone to be reconverged. It is not hard to show from simple order of magnitude estimates that, for sufficiently large collections of mass, the criterion can indeed be satisfied before densities or curvatures became excessive, and without any assumption of symmetry.

Once this criterion has been satisfied, then according to a precise theorem in general relativity put forward by Stephen Hawking and myself, it follows that there must be a space time singularity somewhere. The theorem does not say that this singularity is necessarily of the same character as that encountered in the centre of a spherically symmetrical black hole, but it is hard to avoid the inference that tidal effects which approach infinity will occur, producing a region of space time where infinitely strong gravitational forces literally squeeze matter and photons out of existence.

Physicists are unhappy with a theory that predicts the evolution of such a truly physical singular state. In the past whenever a singularity was encountered in a theory, it was generally a warning that the theory in its present form was breaking down and new theoretical tools were needed. In the case of black holes we theoreticians are again being presented with a situation of this kind, but one more serious than before, because here the singularity refers to the very structure of space and time.

There are two distinct possibilities at this stage. It may be that the resulting singularity is such that signals can escape from it which can be observed at large distances. This is the more alarming of the two possibilities and it is also the more conjectural. Such a singularity is called 'naked'. The possibility of naked singularities is alarming because the physical effects of near-infinite space time curvatures are quite unknown. If these effects can influence the outside world, then an essential uncertainty is introduced into present physical theory.

On the other hand it is possible that the singularities resulting from gravitational collapse are always hidden from view, as was the case in the spherically symmetrical situation considered above. In that case this essential uncertainty will not arise. This is the hypothesis of 'Cosmic Censorship', according to which naked singularities are forbidden, each singularity being necessarily clothed by an absolute event horizon. There is perhaps some slight theoretical evidence in favour of Cosmic Censorship, but it is only very slight. After all, the consequences of the big bang singularity were decidedly *not* hidden; here was the biggest naked singularity of all time. However, it would be inaccurate to think of the big bang as a violation of Cosmic Censorship. We are concerned here only with singularities which arise in the collapse of perfectly reasonable nonsingular matter. I would personally tend to believe that in situations which do not differ too much initially from that of spherical symmetry, the Cosmic Censorship principle is valid. But in more extreme cases the question is, to my mind, quite open. Perhaps there is even observational evidence against Cosmic Censorship. This is a matter to which I shall return later.

If we assume the Cosmic Censorship hypothesis is true, then once the focusing criterion is satisfied an absolute event horizon must arise. This horizon will have a well-defined cross sectional area which will have a tendency to increase with time (black holes can grow but never shrink) but it seems reasonable to suppose that a black hole, left to itself, will settle down to a stationary state. However, we must be cautious in our use of intuition. It might also seem reasonable to suppose that given the vast range of structures, configurations and complexities of the bodies which could have collapsed to a black hole in the first place, the configuration of the black hole itself could also be complex. Some remarkable work by Werner Israel, Brandon Carter and Stephen Hawking has shown that this is probably not the case. Only a very restricted class of stationary black hole configurations can arise. They are uniquely characterised by the value of the mass, spin and charge of the hole. Einstein's equations for the general theory of relativity have been solved explicitly for this problem by Roy P. Kerr and the solution generalised to include charge by Ezra Newman and his co-workers. The reason the asymmetries present in the collapsing body do not show up in the final state of the black hole is that once the hole is formed the body that produced it has little influence on the hole's subsequent behaviour. The black hole is best thought of as a self-sustaining gravitational field governed by the internal non-linear dynamics of the general theory of relativity. These dynamics allow the asymmetries in the gravitational field of the hole to be carried away in the form of gravitational waves as the hole settles down into a stable configuration.

We have seen that a material object, once swallowed by a black hole, cannot escape. On the other hand, there are mechanisms whereby some of the energy content of the black hole can be extracted. One such mechanism involves the coalescing of two black holes. This process would be accompanied by the copious emission of gravitational waves, whose total energy should be a substantial fraction of the initial rest mass energy of the black holes. Another mechanism would be to allow a particle to fall into a region close to the event horizon of a rotating black hole. The particle splits into two particles in such a way that one falls into the hole and the other escapes back to infinity with *more* mass energy than the initial particle had. In this way rotational energy of the black hole is transferred to the particle motions outside the hole. In the process the black hole loses mass and spin. In principle, this is an extremely efficient means of converting rest-mass to energy—much more efficient than nuclear fission or fusion! But there is an absolute limit even to this procedure. In the most extreme case the mass of the black hole might conceivably be reduced down to 0.707 of its original value by this sort of general procedure. In any case it is hard to envisage this process being effective in an actual astrophysical situation.

Let us now consider the situation inside the black hole, and the general relativistic implication of the existence of a space-time singularity. Since a 'singularity' means a region of breakdown of physical theory, we have the curious situation that, here, general relativity is predicting its own downfall! But perhaps we should not be too surprised at this; after all we are treating general relativity only in its capacity as a classical theory. When the curvature of space time becomes enormous, quantum effects must eventually play a dominant role. When the radius of space-time curvature becomes as small as, say (10^{-13}) cm (roughly the radius of an elementary particle) then the theory of particle physics as understood at present, must break down. If the radius of space-time curvature ever becomes as small as (10^{-33}) cm (and the implications of what we have said so far are that it will be that small somewhere inside a black hole—unless theory breaks down before this), then we cannot avoid having to apply quantum mechanics to the structure of space-time itself. At present there is no satisfactory theory for doing this. It should be emphasised that there is no reason to believe that a new theory is needed to deal with situations less extreme than these. Only *at* the singularities would we have trouble. And if Cosmic Censorship holds true, the absolute event horizon would prevent the effects of such new physics having any influence on the outside world.

A question which often puzzles people in connection with black holes goes roughly as follows: If the absolute event horizon is so effective in shielding the contents of the black hole from outside, how is it that the black hole can still exert a gravitational influence on other bodies? How can it be that the gravitational field of the collapsed body escapes the black hole even though no information or signal can get out? In fact it is not really accurate to say that the gravitational field 'escapes'. It would be more true to say that the gravitational field is basically that of the body *before* collapse. If subsequently the body is annihilated at the centre of the hole, the exterior gravitational field cannot cut off. It would be *this* which would require information to escape the hole, because the gravitational field would have to 'know' when the body producing it had disappeared. So the exterior field does not reflect any change which takes place in the interior. After the body has collapsed in, it is better to think of the black hole as a self-sustaining gravitational field in its own right. It has no further use for the body which originally built it!

Let us return, now, to the observational status of black holes. The most promising place to look for black holes is in binary systems, such as HDE 226868—Cygnus X-1. As I mentioned earlier there are a number of such systems under consideration, with Cygnus X-1 being presently the most convincing candidate. Apart from the nature of the X-ray emission, the identification of the invisible component as a black hole would come from the estimate of its mass, this mass being detected via its gravitational effect on the visible component of the binary system. When the mass turns out to be considerably too large for a white or black dwarf or a neutron star, then the case for it being a black hole would seem to be a very strong one.

There is another aspect to the role of black holes in observational astronomy. We may compare the situation with that which arose in the case of the neutron stars. For many years astronomers had attempted to detect them by searching for certain effects such as X-ray emission which had been predicted also to occur in connection with neutron stars. However when neutron stars were first detected it was via an effect which was totally unexpected and still not really satisfactorily explained, namely the emission of the rapid and regular sharp pulses of electro-magnetic radiation characteristic of a pulsar. It is quite possible that the detection of other black holes will be by some equally unexpected observational side-effect.

There is no shortage of unexplained phenomena in cosmology today that might conceivably be relevant, from the phenomenal energy output of quasars and radio galaxies, the apparently anomalous red-shifts in the spectra of some quasars and galaxies, the explosions in the centres of some galaxies and the discrepancies in the mass measurements of galaxies, to serious questions that seem to remain even about the spiral-arm structure of normal galaxies. Above all, there is the apparent observation by Joseph Weber of the University of Maryland, of gravitational waves emanating from the centre of our own galaxy. As Professor Martin Rees points out in Chapter 10, if these waves are emitted continuously and in all directions uniformly from the galactic centre, the energy they carry would correspond to the loss of mass from the galaxy of many thousands of solar masses per year. This is in gross conflict with other observations.

So far the theory of black holes has not led to a very convincing explanation of this worrying phenomenon, or any of the others, but these are early days yet. In the case of Weber's observations taken at their face value, the best hope would seem to be some mechanism whereby the waves are strongly beamed in the galactic plane. The sun and the solar system are unusually close to the galactic plane—they lie in it to an accuracy of one part in a thousand. With a beaming angle of something like $1/1,000$ the observational conflicts would be removed. Attempts have been made to explain this kind of beaming on the basis of a fast rotating black hole at the centre of our galaxy, but so far they are not very convincing. An alternative to beaming would be the possibility that the production of the waves from the galactic centre is a comparatively short-lived phenomenon, lasting for considerably less than a one-hundredth part of the time that our galaxy has been in existence. (This would be a rather uneasy coincidence, since our appearance on this earth presumably need not have occurred just at the time of extreme gravitational activity at the centre of the galaxy.) Another possibility is that there might be some source much nearer to us lying approximately on the straight line containing us and the galactic centre. However, there are objections to this idea too. Of course there is also the possibility that the observations are in some way spurious and that a phenomenon other than gravitational waves is contriving to produce the effects that Weber observes. A number of other groups have now built detectors and it is to be hoped that this uncertainty will be cleared up in a few years.

Let us assume, for the purposes of argument at least, that Weber's results are substantiated; then the question arises as to how nature produces such waves. Although black holes have usually been suggested as the most likely objects which could be responsible, no plausible detailed explanation seems to be forthcoming. I feel that we should not close our minds to the possibility that some other type of gravitational catastrophe might be occurring at the centre of our galaxy. The question of naked singularities has been largely left aside; but as I have said earlier, there is no very convincing theoretical argument in favour of Cosmic Censorship. Could it be that naked singularities *are* in some way responsible? I think that one should not ignore this possibility, particularly since it offers some hope of producing the beaming necessary to fit in with the observations.

Let me present the following very hypothetical picture. Suppose that some form of large rapidly rotating mass at one time collected at the centre of our galaxy and that this mass approached a regime at which the effects of general relativity dominated over all more conventional forces. Suppose also (and this is the big assumption) that such rotating agglomerations have a tendency to settle into a configuration whose exterior field closely approximates a Kerr solution of the Einstein field equations. Recall that black holes do just this—but I am not now thinking of a black hole! If the rotation remains too great, the solution will not in fact describe a black hole. Instead the matter would contract down until a naked singularity is revealed in a ring around the equatorial region. This of course is very speculative but the model possesses one feature which is perhaps suggestive: the naked singularity would be visible only from the equatorial plane! (This follows from some work by Brandon Carter.) Thus any signal originating at or near the singularity would necessarily be beamed very closely in one plane. If we assume a tie-up between this object and the structure of the galaxy then it is not unreasonable to suppose that the beaming plane and the galactic plane coincide. Finally, since ultra-strong gravitational tidal fields would exist near the singularity one would expect that it would be gravitational waves which would be so beamed. Weber's results might conceivably be explained in this way.

It has often been argued that if naked singularities arise then this situation would be disastrous for physics. I do not share such feelings. True we have, as yet, no theory which can cope with space-time singularities. But I am an optimist. I believe that eventually such a theory will be found. In any case we have for some time been confronted by the profound theoretical problems of the big bang singularity, and a theory is needed to cope with this. Would it not be far more exciting if in addition there were other space-time singularities accessible to view *now* which could supply observational means of testing such a new theory? Perhaps then the mysteries of the initial creation could be more readily comprehended.

The mere facts of modern cosmological science are, as the previous chapters have shown, so startling that anyone wishing to go further and speculate very quickly finds himself right out on a limb. Yet the peculiar observations now being made seem to necessitate speculation. What are the quasars? From whence do they derive their stupendous power? How far away are they? Here in one area alone attempts to provide adequate explanations find the more sober-sided conservative approach wanting. The years to come will bring more penetrating observations and more rigorous analyses, but will the result be elucidation or will we reach, at last, the limits of human understanding?

In the following key chapter Martin Rees, Plumian Professor of Astronomy and Experimental Philosophy, Cambridge University, takes a look not only at the future of the universe, but of cosmology itself. L. H. J.

THE FAR FUTURE
Professor
Martin Rees

Ever since prehistoric times men have wondered about the nature of the cosmos—its origin, its structure, and its ultimate fate. But only in the last fifty years or so has cosmology become a genuine science, in the sense of being an intellectual activity where one tries to formulate theories, and then tests the consequences of these theories against the results of observations and experiments. The present generation of cosmologists is fortunate to be living through an era when millennia of speculation are being supplemented by some tentative glimmerings of, we hope, genuine understanding. The subject has already progressed far enough to enable us to sketch out the large-scale structure of the universe with at least moderate confidence that we are probably on the right lines, and also to say something about the history of the universe and how it has changed.

The most striking result of observational cosmology is that the universe—in its gross structure—seems to be simpler than one had any right to expect. Provided that we ignore density inhomogeneities on scales up to 100 million light years (which is still small compared with distances of several thousand million light years to which observations now extend) the Hubble law seems to be obeyed with high accuracy, and the overall dynamics are very well described by the simplest conceivable idealisation, in which the universe is assumed uniform and the expansion isotropic.

By synthesising the disparate strands of evidence bearing on the problem, most cosmologists have reached a tentative consensus of how—in broad outline—the universe might have evolved to its present state. To recap on what the authors of previous chapters have described (*pace* Professor

Chapter 10

Narlikar), about 10,000 million years ago, all the material in the universe—all the stuff of which galaxies are now composed—constituted an exceedingly compressed and hot gas (hotter, in fact, than the centre of the sun). The intense radiation in this fireball, though cooled and diluted by the expansion, would still be around, pervading the whole universe: this is the interpretation of the microwave background radiation discovered by Penzias and Wilson—an 'echo', as it were, of the 'explosion' which initiated the universal expansion. The early universe would not have been *completely* smooth and homogeneous—it may even have been turbulent—and the primordial irregularities eventually developed into galaxies. It may, however, have taken 1,000 million (10^9) years (10 per cent of the present 'age of the universe') for these fluctuations to condense out into gravitationally bound systems. The most distant radio sources, and maybe the quasars as well, are so far away that the radiation now reaching us has been travelling towards us for 80 or even 90 per cent of the time elapsed since the initial big bang. That is to say, it began its journey when the universe was only 10 or 20 per cent of its present age. The radio and optical evidence suggests that the universe would have appeared much more violent and active at these early epochs. The cosmic radio background would have been 100 times more intense; and an astronomer who observed at that time would have found his nearest quasar perhaps fifty times closer to him than is the case in these relatively quiescent times. Perhaps galaxies were more prone to indulge in radio outbursts—or to flare up into quasars—when they were young.

The microwave background is a 'fossil' from even earlier stages in the expansion. There seems to be another important vestige of the primordial fireball in the present-day universe: the element helium, which comprises about a quarter of the mass of most stars, including the sun. The other chemical elements are believed to have been synthesised by nuclear processes in stars, but it proved hard to understand how stars could have synthesised all the helium. So it was gratifying both to cosmologists and to experts on nucleogenesis when the expected composition of material emerging from the big bang was calculated and found to be about 75 per cent hydrogen and 25 per cent helium. The synthesis of helium would have occurred within only a *few minutes* of the big bang! This epitomises how the discovery of the microwave background has extended the scope of cosmology by bringing remote eras that were previously speculative within the scope of quantitative scientific discussion.

The present data in cosmology are still limited, ambiguous and fragmentary; and they all depend on complex instruments stretched right to the limits of their sensitivity and performance. Therefore, even if this general picture seems consistent with what is known at the moment, it would be rash to bet *too* heavily on it being correct, even in outline. Moreover, self-consistency is, of course, no guarantee of truth in itself. Many more observations are essential before we can state with real confidence whether the real overall structure of the physical universe has been revealed to us; or whether our current ideas will eventually be discarded and superseded as surely as were Ptolemy's epicycles.

The consistency of most of the available data with this general picture encourages most cosmologists to adopt the hot big bang theory as a working hypothesis, which should form the basis for interpreting new observations until some better theory emerges, or until some glaring contradiction reveals itself. There are some, however, who believe that there are *already* a few hard facts which cannot be incorporated into this accepted scheme; that conventional physics (which, in this context, includes Einstein's general theory of relativity) is inadequate to explain what we already know, and that some radical rethinking is necessary.

The current debate between the 'iconoclasts' and the 'conservatives' centres on the vexed question of quasar red-shifts. While nobody disputes that the red-shifts of ordinary galaxies arise from the universal expansion (according to Hubble's law) a small but vocal group of astronomers has—ever since quasars were discovered—suggested that the quasar red-shifts may somehow be special. Were this really the case, it would mean that the quasars could be much closer to us than is customarily assumed, and their typical power output correspondingly smaller. But adoption of this view also entails finding some other interpretation of the red-shift.

As even the most orthodox upholder of the conventional wisdom must admit, the evidence that quasars do indeed lie at great distances is fairly circumstantial. Attempts to plot a magnitude red-shift relation for the 250 or so quasars with known red-shift yield essentially a scatter diagram: there is a broad spread in the range of apparent magnitudes at a given red-shift, and no obvious trend or correlation is manifest. Conservatives would say that this merely indicates that quasars have a wide range of intrinsic luminosities (and indeed the fact that some quasars have been seen to vary by more than a magnitude in only a few months certainly precludes their being good standard candles, on any theory)! And it is possible that some evolutionary effect according to which the more remote quasars are intrinsically brighter masks the magnitude red-shift correlation that one would otherwise expect to find. Had quasars been discovered *before* galaxies, however, the Hubble law would certainly have been less readily accepted, and proposals that red-shifts could be due to something other than the expansion of the universe would have been accorded a more sympathetic reception.

Support for the cosmological interpretation of quasar red-shifts comes from radio observations which show that the apparent radio sizes of quasars are inversely correlated with red-shifts (as is expected if the red-shifts are indeed distance-indicators). Moreover the linear sizes inferred when the distance is derived from Hubble's law are more or less the same as the inferred sizes of radio sources associated with ordinary galaxies whose distance is not in question. If, therefore, the quasars are actually closer, by a certain factor, than their red-shifts suggest, then their radio dimensions must be smaller than those of radio galaxies by about the same factor. This would seem a rather artificial and contrived situation.

A few other circumstantial arguments supporting the cosmological interpretation are worth mentioning. Careful analysis of the magnitude red-shift relation actually reveals that it is not a pure scatter diagram, but that the magnitude of the *brightest* object in a given red-shift range is indeed somewhat correlated with red-shift. Also, a very interesting study by Allan Sandage of the colours of quasars reveals the following suggestive effect. Those quasars which, on the cosmological hypothesis, have the lowest intrinsic luminosity, tend to be less blue than the typical quasar. This is precisely what one would expect if quasars are ultra-bright galactic nuclei—the (redder) light from the underlying galaxy would appreciably affect the observed colours when the quasar has a relatively low luminosity; whereas the blue light from the most powerful quasars would completely swamp the ordinary galactic contribution. Another—rather weak—argument which suggests that quasars cannot be too local follows from the same considerations as underlie Olbers' paradox. If quasars are indeed cosmological, we are in effect resolving individual objects out to the edge of the observable universe; if, contrariwise, the observed objects are relatively nearby, the much larger number of more distant ones too faint to be detected individually could contribute a strong integrated background. The fact that known sources contribute about half of the total brightness of the radio sky, therefore, sets a lower limit to the distance of a typical quasar.

Ever since their initial discovery, it has been recognised that quasars—and in particular the colossal and very concentrated energy source which must be involved—pose extreme theoretical problems. The most popular theories involve either one superstar—a single object millions of times more massive than the sun—or else a cluster of stars so densely packed that stellar collisions are frequent, or where all the stars somehow evolve more or less in unison to the supernova stage. One motivation for adopting the local hypothesis was indeed that this would reduce the total energies, though at the cost of invoking a new red-shift mechanism. The red-shift has been explained, for example, by saying that all observed quasars may have been flung out from the centre of our galaxy, or perhaps a neighbouring galaxy, with velocities close to the speed of light. They would then display a doppler effect having nothing to do with the expansion of the universe. Adherents of this view are forced, however, to invoke a hypothetical explosion with properties fully as mysterious as the quasars themselves. Why, also, do we see no blue-shifted quasars which are coming towards us instead of going away? Another possibility is that quasars display a so-called Einstein red-shift—an effect which would be large if they were so massive, or so compact, that the light escaping from them lost a substantial fraction of its energy climbing out of the gravitational field. Another possibility is that the red-shifts may be 'metaphysical' in the sense that they involve some as yet unenvisioned new law of nature.

Over the last few years, much effort has been expended in efforts to devise plausible models for quasars based on these rival interpretations of the red-shift. The difficulties with the 'local doppler' and Einstein red-shift hypotheses seem, as a result, to have become more serious. This may, however, reflect merely the limited ingenuity of the astrophysicists who have addressed themselves to the problem, and can therefore never be a very convincing line of argument against the local hypothesis itself. On the other hand, observers have accumulated much more evidence for a continuity of properties between quasars and various kinds of galaxies whose central regions seem to have undergone a violent explosion. This suggests that the quasars may be extreme cases of this explosive behaviour.

The most direct way of finding out how far away the quasars lie would be by finding one which is obviously physically associated with another object—a galaxy or a cluster of galaxies, for instance—with a bona fide red-shift obeying the Hubble law. If the quasars are cosmological, this should only be possible for the ones with relatively small red-shifts—the others then being so far away that no ordinary galaxy at the same distance would be bright enough to be seen. Two or three quasars with small red-shifts, however, do appear to be in the same direction as clusters of galaxies with more or less the same red-shift. This suggests, of course, that these quasars (at least) really do lie in clusters and that their red-shifts are in no way anomalous. One must set against this, however, various curious cases studied by Halton Arp, Geoffrey Burbidge and others, where quasars appear to be physically linked by a luminous bridge to a fairly nearby galaxy (with negligibly small red-shift). If this really indicated a genuine physical connection, then obviously the quasars' large red-shift could not be a valid indication of their distance. Although many associations of this type have been discussed in the literature, it is hard to assess whether they are genuine, or whether they are simply due to the chance alignment of a foreground galaxy with a distant quasar. One should in principle be able to apply statistical tests to check whether these phenomena occur more often than can be explained by chance.

However, in assessing how improbable a particular correlation is one must take account of all the other possibilities that might equally well have occurred. For instance, when there is said to be only one chance in 200 of getting, by pure coincidence, so many close apparent associations between a particular sample of galaxies and a particular type of quasar, we should not blindly accept this as statistically significant without applying a discount to allow for all the other correlations, involving differently chosen samples, which are *not* observed but which are no more improbable. It is all too easy to see patterns in random sets of points. The crucial test is whether a hypothesis has predictive power, and applies not just to the objects in which the alleged effect was first noticed, but to some new samples as well.

Another problem with evaluating the significance of the various anomalies claimed by Arp is that his objects were generally singled out for study only because they looked peculiar or were near to a peculiar object (plate 8). One would like to know how many ordinary galaxies would reveal bridges emerging from them (but *not* ending on a quasar) if subjected to comparably intensive scrutiny. Despite these doubts, however, Arp has certainly produced enough evidence to call into question all conventional ideas about galaxies and quasars, and make other astronomers feel uneasy or elated, according to temperament. One hopes that further work, on samples of objects chosen by predetermined criteria, will eventually settle this fundamental question.

The question of whether quasars are near or far is reminiscent of an astronomical controversy which took place nearly 200 years ago concerning the reality of binary stars. Many instances were known where pairs of stars lay close together on the sky, and John Mitchell (a clergyman and amateur astronomer) showed statistically that there were too many such pairs for them to be merely chance superpositions of foreground and background stars. He therefore argued that these stars must be physically associated 'either by gravity . . . or by some other law or appointment of the Creator.' One of the leading astronomers of the time, William Herschel, contested this conclusion on the grounds that the pairs of stars frequently had very different apparent magnitudes. Since it was well known (though, as we now realise, completely false!) that all stars had the same intrinsic luminosity, their magnitudes would be a measure of their distances, so the members of the alleged binary pairs must, he claimed, be at very different distances from us. It took 36 years for Herschel to change his mind. We readily see a parallel between this debate and the current controversy between those who believe that the redshifts of quasars are a true measure of their distance, and those who adduce statistical evidence to the contrary. But we cannot predict whether the upholders of orthodoxy will again be forced to recant. Perhaps some quasars are local and others are not! This would really land us with the worst of both worlds, but would provide an amusing parallel with the outcome of the debate early this century concerning the nature of diffuse nebulae.

Let us hope that the present dispute will be settled more rapidly than its eighteenth-century counterpart. However, one must remember that most of our knowledge of cosmology comes from studying very faint objects. It is a slow procedure to obtain good spectra of such objects, and their angular sizes are typically so minute that there is little possibility of learning much about their spatial structure. Therefore, barring some dramatic and unforeseen discovery, progress is unlikely to be rapid because of the limited number of suitable instruments and of skilled observers.

This situation contrasts markedly with that elsewhere in astronomy—for instance, the study of interstellar molecules, and of radio (and now X-ray) pulsars. Neither of these areas of research existed when quasars were discovered, but already an almost embarrassing quantity of information has been amassed. Though this information has yet to yield a proportionate amount of true knowledge and understanding, analysis of these comparatively local phenomena within our own galaxy will undoubtedly distract many theorists' attention from the cosmological problem in the next few years. The unsuspected complexity of the molecules found in interstellar clouds has initiated the new subject of interstellar chemistry, which will help us to understand the processes whereby stars form from tenuous interstellar gas, and perhaps even provide some clues to the origin of life.

Pulsars are the supreme example of how the universe in effect constitutes a laboratory where matter can be observed under extreme conditions of density, pressure and temperature that could never be simulated artificially, and where the known laws of physics can be tested perhaps to breaking point. Pulsars are believed to be spinning neutron stars—objects as massive as the sun, but so compressed that they are only a few miles across, the internal densities being so high that the whole earth, at the same density, would fit inside St Paul's Cathedral. Neutron star matter would display extraordinary properties analogous to very low temperature liquids in the laboratory. Pulsars also possess magnetic fields millions of times stronger than can be generated on the earth; and their characteristic pulses represent a concentration of radiant energy far exceeding that of any laboratory laser. The force of gravity on a neutron star is so great that Newton's theory is inadequate to describe it accurately. Thus study of pulsars may help to decide between general relativity and rival theories of gravitation. Even more exciting, as Professor Penrose has described in Chapter 9, is the possibility that stars may exist which have collapsed even beyond the neutron star state, to become black holes—bodies where gravity has overwhelmed all pressure forces, and is so strong that not even light can escape from them (in other words, their Einstein red-shift is infinite). An object which has undergone complete gravitational collapse still exerts a gravitational pull on surrounding material. Any gas accreted by a black hole would radiate—probably in the X-ray part of the spectrum—as it swirls inward. Some astronomers believe that X-ray sources answering this description have already been found within our galaxy. If so, this opens the way to testing the most crucial and fantastic predictions of Einstein's theory.

These galactic phenomena certainly represent the most exciting short-term prospects in astronomy. They may even allow us to examine, at relatively close range, the same exotic processes that occur on a much larger scale in the quasars and radio galaxies. They could thus help to elucidate the nature of these enigmatic objects, and thereby be of indirect relevance to cosmology. But the nature of quasars will probably only be settled when many more observations have been accumulated, and more statistical studies carried out. If the local theory is eventually discredited, we shall have increased confidence that we understand the general evolutionary history of the universe, quasars being interpreted as hyperactive galactic nuclei. On the other hand, even if some other interpretation of quasar red-shift is forced on us, there will be no reason, necessarily, to abandon the hot big bang theory. However, quasars will then become even more important, because they may involve some basically new law of nature. (After all, a physicist whose laboratory was floating freely in space would probably never have discovered gravity, because this force is very weak unless a large mass such as the earth is involved. Maybe, therefore, there are other effects, insignificant even on the scale of the Solar System, which nonetheless play a crucial role in an object like a quasar).

Although one may well marvel at there being any progress in cosmology, it is as well to recall how extensive and basic our ignorance still is. Professor McCrea has already pointed out in Chapter 7 that we know very little about the large-scale distribution of matter in space—galaxies, clusters of galaxies, perhaps clusters of clusters. And we know next to nothing about how smooth the universe was at early times. We do not know whether the overall dynamics of the expansion are really governed by Einstein's theory of gravitation. We do not know whether continuous creation is going on—although the strict steady state theory is almost certainly wrong, new material may still conceivably emerge from galactic nuclei or other exotic environments. We do not know over what range of space and time the so-called fundamental constants may be treated as such—there is no positive evidence that these basic numbers actually do vary, but to apply the ordinary laws of physics to the *very* early stages of a hot big bang (as is done in calculations of the helium abundance) may well involve an unwarranted extrapolation. And, above all, the large-scale uniformity of the universe—the one circumstance which makes progress in cosmology feasible—still poses a major mystery.

As newer and more sensitive techniques are brought to bear on the subject and more is learnt, the details of the picture may well be filled in until we come to understand how the primordial matter aggregated into galaxies, why galaxies have the shapes and sizes that are observed, and how they evolve. We may even grasp some of the links in the long and mysterious chain of processes whereby some chemicals assembled into such incredibly complex structures as our own brains, endowed with the capacity to contemplate and reflect on the vistas around them. But this progress is certainly not guaranteed—the fact that most of our current knowledge lends itself to a consistent interpretation may reflect the paucity of data rather than the excellence of our theories, and each new observation may turn out to impose new constraints that make interpretation more difficult. One can only hope that we do not merely exchange our current fog of ignorance for a complex and impenetrable maze of confusion.

Having drawn some tentative conclusions about how the universe has evolved from the primordial fireball to its present state, one is tempted to speculate about its future and its eventual fate. The conventional theories distinguish two contrasting scenarios: either the universe will carry on expanding for ever, or else the expansion is slowing down to such an extent that it will eventually stop, and be followed by a recontraction. This question can be tackled observationally in the following manner. The velocity–distance relation for galaxies obviously refers to the velocity of the galaxies at the time the light was emitted. So, if Hubble's work could be extended to galaxies so far away that their light has taken a good fraction of the age of the universe to reach us, then the deceleration should be measurable. A lot of effort has been put into solving this problem in the last decade. So far, however, the results are disappointingly inconclusive, basically because—even with the 200-inch telescope—galaxies become invisibly faint before one gets to distances where the effects of the deceleration should really show up.

Another difficulty is that one needs to know how bright the distant galaxies are intrinsically: galaxies may have been systematically brighter (or fainter) when they were younger; and not enough is known about the evolution of galaxies, and the stellar populations within them, to enable us to make reliable calculations of what correction to apply. The radio astronomer can probe deeper into space (and thus further back into the past) than his optical colleagues. However, one cannot obtain *any* useful estimate of the deceleration from radio data alone, because the evolutionary corrections are larger and even more uncertain than for galaxies. Nor has the discovery of quasars improved the situation: even if one could dismiss all doubts about the true nature of their red-shift, no analysis of the red-shift distribution of quasars can tell us anything about the dynamics of the universe until we know how their average properties evolve.

But there is another—more indirect—way of trying to determine how much the universal expansion is slowing down. Imagine that a big sphere is shattered by an explosion, the debris flying off in all directions. Each fragment feels the gravitational pull of all the others, and this causes the expansion to decelerate. If the explosion were sufficiently violent, then the debris would fly apart for ever; but if the fragments were not moving quite so fast, gravity might bind them together strongly enough to bring the expansion to a halt. The material would then re-collapse. More or less the same argument probably holds for the universe. One might feel somewhat uneasy about applying a result based on Newton's theory of gravity to the whole universe. But even though one cannot describe the global properties of the universe properly (nor the propagation of light) without using a more sophisticated theory such as Einstein's general relativity, the dynamics of the expansion *are* the same as in Newton's theory. So we can rephrase our earlier question as: does the universe have the escape velocity or not?

In the case of the galaxies (which, for the purposes of this argument, are regarded as fragments of the expanding universe) we know the expansion velocity. What we don't know is the amount of gravitating matter which is tending to brake the expansion. It is a straightforward procedure to calculate how much material is needed in order to halt the expansion: it works out at about one atom in each million cubic centimetres. If the average concentration of material were *below* this so-called critical density, we would expect the universe to continue expanding for ever; but if the mean density *exceeded* the critical density, the universe would seem destined eventually to re-contract. There are various ways of estimating the masses of individual galaxies. Also, of course, one knows roughly how many galaxies—how many spirals, how many ellipticals, etc—there are in a typical volume of space. These estimates are bedevilled by many uncertainties, but it looks as though the amount of material in galaxies, if spread uniformly through space, would fall short of the critical density by a factor of at least thirty. At first sight, one might accept this as evidence that the universe will go on expanding for ever; but this inference would really be unjustified, because there may be a lot more material embodied in some form other than ordinary galaxies. Galaxies are, admittedly, the most prominent features in the sky when we look with an optical telescope: but there is no reason to believe that everything in the universe shines. There may be many objects so cool that they radiate predominantly in the infra-red, and *absorb* light instead of emitting it—dead galaxies, for example, whose stars have all exhausted their nuclear energy, or objects shrouded by opaque clouds of dust.

Alternatively, some objects may be so hot that their emission is concentrated in the ultraviolet and X-ray bands of the electromagnetic spectrum. This kind of astronomical work was not feasible until quite recently. This is because the air is very opaque so one must send equipment to high altitudes above the earth's atmosphere, in a rocket or a satellite, to make observations. These new astronomies are still in the same pioneering stage of development that radio astronomy was in twenty years ago, and when we recall how different the radio universe has proved to be from that revealed by optical telescope it would not be surprising if space astronomy were to disclose an equally complex range of unsuspected objects. So our present inventory of the contents of the universe may well prove exceedingly biased and incomplete.

One widely discussed possibility is that there may be a large amount of diffuse gas *between* the galaxies. There is, after all, no reason to expect all or even most of the primordial hydrogen and helium to have condensed into galaxies. Despite intensive efforts to detect emission or absorption by such a gas, no firm evidence for its existence has been found, though intergalactic gas may in fact be responsible for emitting most of the cosmic X-rays reaching us from beyond our own galaxy. And in any case, absence of evidence need not be evidence of absence.

At present, therefore, there is no definite reason for believing that there is enough gravitating material in the universe to bring the expansion to a halt. But it remains quite conceivable that the universe contains a great deal of stuff even more elusive than intergalactic gas. For instance, the 'critical density' could be provided by neutrinos, gravitational waves, or black holes without there being the slightest chance of detection by present techniques.

We, therefore, cannot rule out the possibility that eventually the expansion will stop and turn into a contraction. Distant galaxies, displaying blue-shifts instead of red-shifts, would eventually collide and merge with one another. As the contraction proceeded further, the sky would become brighter and brighter; and eventually all the stars would explode (because the sky would be hotter than the fuel in their interiors!), everything in the universe being finally engulfed in a fireball like that from which, according to most cosmologists, it emerged: the ultimate, universal, gravitational collapse. But there is no immediate cause for concern—our breathing space before this cataclysm should be thousands of millions of years, at the very least. If the universe does not have the 'critical density', then the galaxies will continue to recede from us for ever. Each galaxy would fade to a dull red glow as its constituent stars exhaust their available energy, and the supply of gas from which new bright stars can condense is inexorably depleted.

Unless a large amount of hitherto undetected intergalactic material is discovered, or else the deceleration can be determined from the red-shift magnitude relation for distant galaxies (after applying appropriate evolutionary corrections) we have no way of telling which of these contrasting fates awaits the universe.

The astronomer—in contrast to most other physical scientists—is unable to perform 'experiments' on his subjects of study. But his helplessness in this respect is compensated by the enormous number and variety of stars and galaxies accessible to observation. By studying the relative numbers of stars of different types, and correlating their colour, brightness, etc, he can infer the life history of a typical star—rather as you might infer the life history of a tree (even if you had never seen one before) by a day's observation of a forest containing trees of all ages. Astronomers can also learn something about the structure and evolution of galaxies in general by investigating peculiar galaxies which have undergone some violent explosion, or been distorted by a close collision with a near neighbour: in these cases, nature has performed an experiment for us.

Cosmology, however, is by definition the study of a unique object and a unique event. No physicist would be happy to base a theory on a single unrepeatable phenomenon or experiment, but we plainly cannot check our cosmological ideas by applying them to other universes. Nor can we repeat or rerun the past—though the finite speed of light allows us to sample the past by looking at very remote objects. Cosmology is generally, therefore, regarded as a descriptive science, whose aim is to describe the overall properties and evolution of the cosmos, and to interpret as many features as possible in terms of physical laws known from the results of local experiment. It is normally assumed that matter everywhere, and at all times, behaves in the same way—in other words, that the masses and charges of protons, electrons and other particles, and their various energy levels, are indeed universal constants. This assumption is supported by all measurements of the spectra of distant objects, which seem to imply that atoms behave identically throughout the observable universe. It is often forgotten what a remarkable circumstance this is in itself.

If the so-called hot big bang theory is correct, we can already relate some aspects of the present universe—the cosmic helium abundance, for instance, and perhaps the structure of galaxies and clusters—to processes occurring at very early epochs. However, the further one extrapolates back towards the initial singularity, the less confidence one has that the known laws of physics are either relevant or complete. Theorists differ in the extent to which they are prepared to make such extrapolations. The more cautious among them would regard at least the first million years of the expansion as *terra incognita*; those astrophysicists who believe helium is primordial invoke processes which occurred only a few seconds after the expansion started; and some people have given scholarly and serious discussions of the first microsecond! (Of course, if one measures time on a logarithmic scale, that is probably where all the action is.) But all cosmologists would accept that there is *some* epoch before which physics as we know it can give no useful guidance at all. Maybe some new physical insights into quantum gravity may allow us to push back the frontiers of the unknown a bit further, but some essential limitation on our knowledge will remain.

The restricted character of cosmological explanation can perhaps be clarified by a rather frivolous illustration. One very early cosmology—due, I think, to the Indians—was as follows: the earth, which is more or less flat, is supported by four elephants, one at each corner. The elephants stand on the back of a turtle which is swimming in an ocean. Even when this idea was promulgated, it must have occurred to some people to ask what was at the bottom of the ocean. Presumably there was no satisfactory answer. But the point of this illustration is that the modern cosmologist is in no better shape—he still, eventually, has to fall back on initial conditions, and say 'things are as they are because they were as they were'.

Cosmologists normally regard the laws of microphysics as autonomous, in the sense that these laws govern the gross behaviour of matter throughout space, without being themselves affected by the universe. Some people, however, have come to suspect that this view may be mistaken, and have suggested instead that microphysics and macrophysics—the world of the elementary particle, and the universe as a whole—may somehow be interdependent. In other words, perhaps the structure of the universe determines what atoms look like, as well as vice versa. If this view were the right one, then we could not hope fully to understand what happens within a single atom without taking the global aspects of the cosmos into account, and cosmology would become an integral part of fundamental physics. Of course, adherents of Mach's Principle have long held that the inertial properties of all objects derive from distant matter, but the idea that other constants of particle physics may be cosmologically determined is less widely shared.

It is then tempting to speculate further. If our local physics depends on the structure of the whole universe, then (if one understood the interconnections) one could envisage other universes where matter behaved very differently. The whole character of laboratory physics as we know it, and the properties of everyday objects, are essentially determined by a few basic ratios—the relative masses of protons, electrons and neutrons; the strength of the electrical, nuclear and gravitational forces between them; and the so-called fine structure constant, whose value determines the energy of the quantised orbits of electrons in atoms. The chemical elements—carbon, oxygen, iron, etc—are only stable because of a fairly delicate balance between the electrical and nuclear forces within complex nuclei. If the laws of physics—and, in particular, the relative strengths of nuclear and electrical forces—were changed too much, then neither the chemical elements, nor complex molecules, could exist. Furthermore, if the force of gravity were too different from its actual value, stars as we know them could not exist either.

Among the requirements for our own existence are the occurrence of complex molecules, a warm environment where reactions between these molecules can occur, and the existence of stars in whose cores the heavy elements were synthesised by nuclear reactions. (It is believed that all the heavy elements in our bodies, and throughout the solar system, were manufactured in stars which completed their evolution and—perhaps in a supernova explosion—ejected material back in the interstellar medium before the sun formed. The solar system then condensed from gas contaminated by debris from earlier generations of stars.) Perhaps, then, our own presence somehow necessitates a universe not too different from the one we see around us, in the sense that various strict constraints are preconditions for our existence. So one could argue that any hypothetical universe too different from our own might be unobservable in principle, because the environment within it would not be propitious for the evolution of observers. It is hard, of course, to construct a rigorous argument along these lines, because we may be taking an over-restrictive view of what constitutes an observer. Perhaps some kind of complex conscious entity could exist even in a cold universe devoid of stars and composed only of hydrogen and helium.

If our universe is destined to collapse—and we may, before long, have observational evidence as to whether this is likely or not—it is conceivable that the whole span of cosmic history accessible to us is just one cycle of an infinitely repetitive process. We cannot hope, with our present scanty knowledge, to understand how, in effect, the universe might bounce or rebound between one cycle and the next. Perhaps the naïve idea of time, and even the concept of before and after, breaks down under these extreme conditions. Perhaps matter is not merely reprocessed in each bounce—all complex nuclei being broken down into their elementary constituents—but the laws governing its behaviour are laid down anew in each cycle. One might then introduce the concept of an ensemble of universes and envisage—as suggested by Professor John Wheeler of Princeton—a process of natural selection of physical constants whereby some dynamic cycles of the universe are cognisable whereas others, in which the prevailing physical laws preclude the evolution of observers, are unknowable.

If these speculations have anything in them, they obviously enlarge our conception of the physical world in an almost mind-blowing way. But even if they are basically correct, what is the chance of putting them on a serious quantitative basis? The prospects are not necessarily hopeless. It is, after all, the complexity of a process, not its sheer size, that makes it hard to comprehend. For example, we already understand the inside of the sun better than the interior of the earth. The earth is harder to understand because the temperatures and pressures inside it are less extreme than in the sun, where it is so hot that matter is broken down into its simplest constituents. (For analogous reasons, the tiniest living organism is harder to understand fully than any large-scale inanimate phenomenon.) So perhaps the earliest phases of the primordial fireball—where conditions were surely too extreme to permit any complex structures to survive—will not for ever remain beyond our comprehension. On the other hand, the relevant concepts may prove too complex for our minds to grasp. Progress would then have to await the evolution of a more intelligent species than ourselves.

To be a cosmologist at all requires an act of faith at the outset. It would be perfectly reasonable to argue that, since we live in a corner of the universe that could, for all we know, be quite untypical of the universe as a whole, the physical laws that we have discovered may be quite inapplicable on the grand scale. Fortunately, the opposite view also applies. For what could provide a better, more rigorous test for our understanding of the physical universe than the extreme conditions provided by the universe itself?

In the following chapter, John Taylor, Professor of Mathematics at King's College, London, describes the manner in which cosmology is able to extend the range of fundamental particle physics to the point where we are forced to rethink our concepts of matter and of time itself. L. H. J.

MATTER
BEYOND
THE END
OF ITS
TETHER
Professor
John
Taylor

Matter exists on earth in a bewildering array of forms, from the innumerable grains of sand on the sea-shore to the vast oceans covering four-fifths of the earth's surface; from crystals of ice and glittering diamonds to the colourless, odourless gas we all breathe; from infinitesimal organisms like bacteria or viruses only to be seen by a hundred thousand-fold magnification to the earth itself, all 6,000 billion billion tons of it. This matter exists under an enormous range of conditions, from the cold, eternally frozen polar wastes to the molten interior of the earth, which exists at a temperature of thousands of degrees and the fantastically high pressure of a million pounds per square inch.

Man has even extended these already diverse conditions under which matter normally occurs so as to investigate better whether he understands the way it behaves. He has cooled it down almost to absolute zero, he has heated it up to millions of degrees in the interior of a hydrogen bomb, he has speeded it up so that it is travelling only a thousandth of a per cent slower than light itself, the fastest of all moving things we know. By enlarging the conditions under which matter normally exists the laws which it is supposed to obey are tested ever more stringently. For, as Sir Hermann Bondi points out in the first chapter, it is only by trying to falsify the predictions of science that advance in scientific understanding actually occurs. The method of disproving scientific predictions evidently works best when matter is in the most bizarre situations because it is here that our expectation of its behaviour is most likely to be wrong.

The range of natural or man-made conditions under which matter occurs on earth is very small compared with the possibilities available in the heavens. Matter can be found at almost the lowest densities imaginable in inter-stellar and inter-galactic space or compressed to the limits of our understanding in the centres of white dwarfs—those stars which have settled down to a tranquil old age having burnt all their nuclear fuel. Temperatures can be reached in stellar interiors which are thousands of times greater than those possible here on earth. Pressures are reached in white dwarf interiors which are so high that matter begins to behave in unexpected ways.

Yet an attempt to understand the behaviour of matter in these more extreme situations has only begun to test our scientific laws in the heavenly laboratory. The discovery of pulsars, and their explanation as neutron stars composed only of neutronic matter and which, when rotating, are supposed to generate the regular pulses from which the pulsars took their name (as Professor Lynden-Bell has described in Chapter 4) has allowed us to investigate matter at even greater extremes. For while material in the centre of a white dwarf is so compressed that a pea made of it would weigh more than a truck—the material weighing about a hundred million grammes per cubic centimetre—matter inside a neutron star is a million times heavier. It is conditions completely impossible to duplicate on earth, and the observations of pulsars, particularly of the 'glitches' or sudden variations in the frequency of their pulsating signals, which present a very great challenge to our understanding.

We are still far from the ultimate conditions to which we can subject matter. For if a star is too heavy it cannot settle down gracefully to a lengthy old age like a white dwarf or by the ferocious outburst of a supernova to become a pulsar. Its component parts attract themselves together so strongly, that no force in heaven or earth can prevent these parts from collapsing towards each other at an ever faster rate. Ultimately, the escape velocity on the surface of the in-falling star exceeds that of light, so that the star can never be seen again. What is left is the black hole, which is described by Roger Penrose in Chapter 9.

The infalling star, leaving its signature-tune behind so dramatically, continues to collapse ever further. After a finite time it has shrunk to a very small size, and can only continue shrinking. The matter composing it is inescapably compressed to higher and higher pressures. To explain what happens as the gravitational forces become indefinitely great and crush the star to nothing is the sternest challenge facing science today. Nature abhors both a vacuum and a point, and our present theories which require the whole of a star ultimately to be compressed to nothingness at a point are evidently on the wrong track. Not that it will be easy to obtain direct experimental evidence as to the correctness or otherwise of any suggested resolution of the difficulty. But our theories must be altered to remove this hiatus.

Even here, at the centre of the black hole, we have not reached the ultimate. For if we return to the very beginning of the universe, the big bang which took place about 10 billion years ago, we find just as extreme a situation, but now with direct observable consequences for ourselves, vast reaches of time later. The relics of the big bang are all about us, and have already been discussed, in particular by Dennis Sciama, Martin Rees and Donald Lynden-Bell. The surprisingly large amount of helium present in the universe, and the three-degree background radiation are indicators of the extremes suffered by the universe in its very early stages, say in the first few seconds or so.

If we turn the clock back even more to the first billionth or even trillionth of a second our universe was in an even more violent state. Indeed, to postulate a reply to the question raised by Professor McCrea in Chapter 7, it is possible that the galaxies dotting the heavens now are the product of vast fluctuations occurring in those first instants of time. There may even be other residues of the universe's fiery beginning, because it is here that the laws of science were created and can be tested.

In a similar way, if the recession of the galaxies ultimately ceases and they begin to move back towards each other they will move ever closer together to play in reverse order the actions begun at the beginning. As the galaxies approach they will co-mingle and lose their identity, as will their constituent stars, their constituent atoms, the constituents of these atoms, and so on. The scenario at the end of the world is that of the beginning reversed up to some point. But it may not continue to be so; the laws of science are once again strained to breaking point to explain the further lines of the great drama of the universe.

In this chapter I will try to describe how far present science can give a coherent picture of matter 'beyond the end of its tether' in the neutron star, inside the black hole, and at the very beginning and end of the universe. In attempting this adventure into the unknown it will turn out that knowledge being gained here on earth has very great relevance for its success. But we will see that it is in the heavenly laboratory that the greatest challenges to science are presented.

Let us start investigating matter *in extremis* with the neutron star. As Donald Lynden-Bell has already described,

K

this strange cosmological object arises from a star which is so heavy that after its nuclear fuel has been used up the material constituting it cannot be prevented from collapsing to a state where the outer electrons are pushed into their nuclei. There they combine with the protons in the nuclei, so that matter is then composed mainly of neutrons, the electrically uncharged companions of the protons in the atomic nucleus. Such 'neutronisation' of matter does not occur completely until a very high compression is reached, with densities of nearly a million billion grammes per cubic centimetre. This is nearly a billion times more dense than in the centre of a white dwarf. Matter experiences an enormous range of densities before it reaches the neutron star state.

The centre of a white dwarf already presents an intriguing problem. The compression there is already extremely high (above 10,000 grammes per cubic centimetre), due to the vast weight of material pressing down on all sides. Consequently the atoms are pushed so close together as to cause them to strip each other's electrons from around the central, oppositely charged nuclei. The electrons float around freely, like a gas, leaving the charged nuclei nakedly sensing each other's electrical repulsion. The repulsive forces between the nuclei then force them to form an orderly array or lattice, just as in a crystal. Each nucleus has to keep equidistant from its neighbours, so that the various forces on it can balance out.

Such a lattice structure can be formed by the atoms in ordinary matter at a suitably low temperature. On heating these atoms begin to move around their lattice positions with ever increasing speed till ultimately they break away from the lattice and the material melts. For ordinary matter the melting temperature is of the order of thousands of degrees. Because of the far superior force of repulsion between the naked nuclei in the lattice at the centre of a white dwarf, the melting temperature for matter in such a state may be as high as 100 million degrees. Calculations have shown that the central temperature of white dwarfs is usually below 10 million degrees, so that the material at the centre of such stars may be solid. If it were composed of carbon, a not unlikely event, the rhyme

> Twinkle, twinkle little star,
> How I wonder what you are,
> Up above the world so high
> Like a diamond in the sky

would indeed be very apt.

As the compression increases, the nuclei of the crystal lattice are able to capture the very energetic free electrons

flowing around, so becoming neutron rich. The most
stable nuclei are initially those of iron but above about ten
million grammes per cubic centimetre nuclei of nickel are
created. At higher densities the lattice apparently becomes
one of selenium, then germanium, zinc, then nickel again
(though now richer in neutrons than before), iron again, but
the neutron-rich sort, molybdenum, zirconium, strontium,
and finally krypton at the very highest compression. Only
the first forms of iron and the two forms of nickel are stable
in the laboratory. The other nucleus created in laboratory
conditions is that of selenium, which lives for only three
minutes before ejecting an electron. None of the other
neutron-rich nuclei have ever been created on earth, though
they may be in the future with the very powerful heavy
particle accelerators which are under construction at various
points of the globe. The inside of a neutron star is undoubtedly a nuclear physicist's dream come true.

Above a density of about 500 billion grammes per centimetre the nuclei are so rich in neutrons that they cannot
acquire any more, and further neutrons formed when
electrons fuse with protons begin to move around freely.
They form a neutron fluid flowing inside the crystal lattice
of nuclei. As the density increases this fluid becomes increasingly important till, at about 200 billion grammes per
cubic centimetre, nuclei are no longer present. Any that
were formed would have to contain so many neutrons as
to be unstable and would rapidly decay away. The dis-
appearance of nuclei happens rapidly as the density is
augmented, and all that is left is a neutron fluid with a few
protons and electrons floating in it.

The model of a neutron star which can be built up from
this understanding of matter under extremely high compression is one with an outer mantle with density rising
very rapidly from zero at the surface to about 100 million
grammes per cubic centimetre within a metre. There is then
a solid crystalline crust of successively neutron-rich nuclei
for a further several kilometres, with an inner density of
200 billion grammes per cubic centimetre. Further inside
still is a fluid of neutrons, which has the property of super-
fluidity (zero viscosity) so that the fluid can climb out of a
beaker, as well as conduct heat much more readily than either
silver or copper.

The solid crust is extremely incompressible, with a
strength of up to a billion billion times more rigid than
normal steel. Because of this and the general very high
gravitational field on the surface, the highest mountains
possible on a neutron star are about 20 centimetres. But,

make no mistake, if you were able to climb such a mountain, you would expend more energy than if you climbed Everest! There may also be a solid centre to the star if it is massive enough, when the neutrons will be pressed so strongly against each other that they can be compressed no further. Experiments over the past decade have shown that neutrons and protons have a central region that acts like a little hard sphere forming a repulsive core; it does not appear possible to compress protons or neutrons further than this, with their repulsive cores touching, without a great expenditure of energy. Such highly compressed neutron matter becomes as rigid as the crust in that situation.

I have presented a description of the interior of a neutron star which is based on scientific laws extrapolated far beyond their tested range of validity. The picture I have painted will only begin to ring true if it explains any of the unexpected features of neutron stars. Naturally enough it explains the basic fact of the observed rapidly and very regularly pulsating radiation. A neutron star is so compact— only 20 kilometres or so across—that it can rotate rapidly enough to generate radiation with the observed frequency of 30 times or so per second; any less compact object of about the same mass, such as a white dwarf, would tear itself to bits trying to spin around so fast.

There are various features of the pulsar signals which are even more interesting to explain. These are the sudden speed-ups that occur in the frequency of the signal. The two most famous pulsars, that in the Crab nebula and the one in the constellation Vela, have both undergone such changes, the latter being one hundred times larger than the former. A very natural explanation of the speed-up is that it occurs due to the general slowing down of the neutron star's rotation. It is difficult for the rigid crust of the neutron star to accommodate this retardation, and it will only do so after a great deal of stress has already been built up. This will result in the star's equivalent of an earthquake, naturally enough called a starquake. It is the energy released suddenly during this latter that causes the pulsar signal's sudden speed-up. The factor of one hundred difference between the Vela and Crab pulsar changes can be ascribed to the starquake occurring on the surface in the latter case but in the interior in the former (so the event should be called a corequake). This would only be possible if the Vela pulsar has a solid interior, so it will have to be somewhat heavier than the Crab pulsar. Such predictions are testable, so this is one situation in which the heavenly laboratory gives a glimpse of matter *in extremis*.

The interior of a heavy neutron star would seem to be very stable, due to its great rigidity. Yet the gravitational attraction is so strong, due to the large amount of matter involved, that the addition of a little more mass to such a star causes it to collapse into a black hole. Indeed, as Roger Penrose stresses in Chapter 9, the sizes of a neutron star and black hole are not very dissimilar. A neutron star somewhat heavier than the sun, say 40 per cent heavier, has a radius of about 10 kilometres, while a black hole 50 per cent more massive than the sun is about one-third of that size. The central region of a neutron star is expected to be the first to disappear by collapsing in on itself, and such a central black hole will eat its way steadily through the rest of the star.

The problem of matter collapsing into the centre of a black hole has already been described as one of the most difficult, perhaps the hardest, of any facing science today. That may be overstating the case a little, but there is undoubtedly a severe problem. For inside a black hole space and time are essentially interchanged. Instead of the incessant ever-onward flow of time, the 'never-ending stream', as it has been so aptly described here in our normal world, in the black hole there is an inescapable motion to the centre. There is no way to escape this one-way flow. Indeed the harder a spaceman would work on the star's infalling surface to avoid the horror of being crushed out of existence as it collapses to a point, the more energy he concentrates around himself, and so the greater his mass. He thus causes an even greater local collapse to ensue, due to the increased attraction being set up. A similar inevitability arises if the spaceman enters the black hole surface after the star has collapsed beneath it; he cannot escape falling to the centre within a finite time.

The reverse phenomenon to this process of collapse of matter to a point is that of its expansion from a point at some specific time in the past. But this is exactly the one facing any scientific analysis of the very early stage of the universe, immediately after the big bang explosion. The evidence in favour of such a scenario for the beginning of the universe has already been summarised by Dennis Sciama; the most important are the observed 3 degree background radiation and the abundance of helium in older stars. We can thus kill two birds with one stone by considering the various stages which occur in the first few minutes after the big bang. Our discussion will also apply to the very final stage of the universe if it is one of ultimate collapse.

Let us turn back the clock so that we can observe the beginning of the universe in reverse. As we approach ever closer to that event the galaxies become overlapping and eventually lose their identity, as do the stars, and so on. This increasing compression of matter would have the same affect on it as occurs inside a neutron star except that as the beginning is reached the temperature of the universe will rise very rapidly. We live at present in a matter dominated world, but when the universe was much younger and smaller the energy of radiation would have been larger than that of the matter present in it. This is due, essentially, to the fact that light emitted during the present expansion phase of the universe has lost energy fighting its way out against the motion of its source. This gives a shift of frequency of the light—the so-called Doppler red-shift—since energy and frequency are proportional for light. In the very early stages, or very close to the end in a final collapse of all the galaxies, this reduction of frequency and energy is reversed; a blue-shift occurs and the light gains energy.

At an early enough time, roughly about 10,000 years after the big bang, radiation and matter had about equal energy. Before then radiation was dominant, and indeed was so powerful that for the first thousand years all atoms were ionised. From then on radiation developed independently of matter, and the expansion of the universe cooled the background radiation to its present 3 degree temperature. At even earlier times, say about 100 seconds after the beginning, the primitive hydrogen nuclei (protons) still had enough energy to fuse together to produce a suitable number of helium atoms by the process first suggested by the American physicist George Gamow, as Dennis Sciama described in Chapter 5. Helium is composed of two protons and two neutrons, and in order to produce the correct amount of helium it was necessary to have a suitable number of neutrons available along with the protons.

These neutrons can be seen to arise by the process suggested by the Japanese physicist Hayashi in 1950. He pointed out that during about the first second after the big bang there was so much energy available that the radiation present would create many pairs of electrons and positrons. These are particles with the same mass but equal and opposite charge, so they could indeed be created (and have been created on earth) by energetic enough radiation. Protons and electrons can interact to produce neutrons, as can neutrons and positrons to produce protons, so that a balance would have been reached between the number of neutrons and protons. However, after the first second the energy would have dropped so much that the concentration of electron-positron pairs was drastically reduced. The ratio of neutrons to protons was then frozen in for the next several hundred seconds till the neutrons had decayed. But before they did that they would have fused to produce helium, and the resultant proportion of about 10 per cent helium agrees well with observation.

We can trace back even earlier, before the first second. Matter then consisted of neutrons and protons as well as the soon to be extinguished electron-positron pairs. These, along with light radiation and neutrinos (particles of zero charge and mass involved in the transformation between neutrons and protons brought about by the electrons and positrons) is what the universe consisted of. This phase of the universe is comparatively simple to visualise if it is assumed that the various constituents are all in thermal equilibrium—an assumption which, after all, underlies the calculations giving the value of the helium abundance quoted above. Abundances of other elements cannot be used to probe back to this early stage of the universe, because the elements heavier than helium were created inside stars. Small proportions of them were made along with helium in the first few hundred seconds, but they were in such small amounts that they can be neglected.

This equilibrium mass of neutrons, protons, neutrinos, electron–positron pairs and electromagnetic radiation was formed about one ten thousandth of a second after the very beginning, and, supposing our argument to be correct, it would have left at least one definite signature behind itself consisting of the neutrinos decoupled from matter at the end of the first second. Unfortunately this background neutrino radiation will be extremely difficult for us to detect because of the very weak interaction of neutrinos with matter in earth-based detectors. It is only possible to observe high energy neutrinos with any hope of success since they have a more powerful interaction with matter than their less energetic brethren. These have been observed on earth, but the very low energy neutrinos which would be the cooled relic of the primeval neutrinos would be practically impossible to detect.

We can attempt to probe to even earlier times than the first ten thousandth of a second. Before then, particle and radiation energies were so high that a completely new realm has to be entered. Here the strong forces holding the sub-nuclear particles—the proton and neutron—together become dominant. The particles of light—the photons—are thought to be responsible for the Coulomb force between electrically charged particles. It is by exchange of photons with the nucleus that the electron is held inside the atom. In

1935 the Japanese physicist Hideko Yukawa argued by analogy that the strong nuclear forces might be generated by the exchange of a similar but much heavier particle, which he called the meson. (The argument being that the exchange of a heavy particle produces a force which is rapidly damped out, due to the difficulty of transmitting the heavy particle very far.) Since the strong forces only act over a very short distance this meson was required to have a mass of about one-sixth that of the proton. A careful search for such a particle was finally successful in 1947 in the tracks left in photographic plates exposed above the atmosphere, and mesons have since been created artificially on earth in energetic collisions between protons.

Alongside the simple meson of Yukawa a whole host of excited companions have been discovered, all of them inter-acting with each other as strongly as does the neutron with the proton. They have been assigned names derived from letters of the Greek alphabet, the original meson being the π-meson, and further ones being called η, φ, ω, ρ and so on. The neutron and proton have also been found to have excited companions, now denoted by Σ, Λ, Ξ, etc. All of these particles are called hadrons. It is these particles which are produced copiously in the first ten thousandth of a second, and this period of the universe is consequently named the hadron era.

In order to understand in more detail what happened during the hadron era, it is necessary to find out what happens to matter as it is heated up. If energy is supplied to water it ultimately boils away; during its boiling its temperature does not rise even though it is absorbing heat. Before the boiling starts, however, the water temperature increases to boiling point. When we turn the clock back on the universe during the hadron era and the energy of the particles increases we need to assess whether it is easier for boiling to occur with no rise in temperature or on the other hand for the hadrons to acquire increased thermal energy, so that the temperature of the universe rises. The first alternative will give a warm big bang, the second a hot one.

The thermal properties of hadron matter are determined by the number of hadrons that can exist with given mass. If their numbers increase sufficiently fast as the total energy of the system is augmented then it is easier for the system to absorb this increased energy by making more and heavier hadrons than by heating up those already present. In the former case such a break-up of existing hadrons into heavier ones can take place very rapidly due to the strong forces between the particles. So the nature of the big bang—was it hot or was it just warm—is answered by the spectrum of hadrons of ever-increasing mass, in other words by the number of hadrons of a given mass. It is this number which is being determined here on earth, over a limited range of masses, the aim being to describe the universe only after the first ten billionth of a second. So far analysis in detail has revealed the state of affairs down to a time one hundred times longer than that. Present evidence indicates that there is indeed a boiling off of hadrons, and that for considerable periods of time there was a constant temperature of about one and a half million million degrees.

Turning the clock ever further back, it is helpful to use not the usual time as a measure of progress but its logarithm to the base of ten, which is increasingly negative for fractions of a second. I will denote this by 'time'. With this parameter we find that the initial stage of the universe moves back to a 'time' of minus infinity, in other words, it was always 'there'. The hadron era had ceased at 'time' minus four, and the electron-positron pairs were 'frozen in' at a 'time' of zero. Radiation was dissociated from matter at a 'time' of ten, and we are now at 'time' seventeen.

If we go back in 'time' to —23, the first 'jiffy' as it is sometimes called, we reach the situation in which the universe is compressed inside the hadrons of lowest mass, those being the particles with greatest extension in space. This means that these lowest mass states cannot be involved in the dynamics of the universe. It is very difficult for any physical entity having too wide an extension in space to interact with anything. The argument is, of course, a quantum mechanical one, but in graphic terms, if two hadrons of low mass touched, there would not be time for the information about this encounter to pervade the whole of each entity before they flew apart again. Only more massive particles are allowed to be used in constructing the universe. As 'time' becomes more negative there are fewer possible hadron states, and the universe will heat up.

We may go back to even more negative values of 'time', still expecting hadrons to play the dominant role, but with ever fewer of them as the universe shrinks unceasingly. Before the 'time' of about —43 something new can happen, for the hadron era may not have started. The force prominent before then may be that of gravity, its strength being enhanced by the very high temperatures or densities then current, boosting it even above the strong force between hadrons. But the aspect of gravity that is needed during this period is the quantum mechanical one, and here we reach a barrier to present understanding of the material world.

L

Matter under all the conditions we have been able to subject it to here on earth is known to have an underlying probabilistic description. Particles have been found to have a wave-like description, as careful experiments have demonstrated. It is rather strange that the undeniably solid matter on which we rely for strength, such as the chair I am sitting on, is composed of wave motions of some underlying entity. In normal situations the vibrations are spaced very closely together so that the solidity of the matter around us returns. But as I have already hinted in my description of the less massive hadrons, there are situations involving very small objects such as the sub-nuclear particles, where the wave aspects of matter are supreme.

We have to accept, then, that the entities out of which all matter is built can best be described as waves, and these in fact are waves of probability. It is not possible to state that a particle is at a point with certainty, but only that if it is observed in many copies of a particular situation it will be at the point in question a certain proportion of times. This probabilistic description is expressed quantitatively by means of a wave function for the system under investigation. This function contains all the possible experimental results which can be obtained by measuring the system, which implies that these results can only be gathered by investigating many copies of the system.

Quantum mechanics, or wave mechanics is the mechanics which is supposed to govern the interaction between pieces of matter when they are described by wave functions. It replaces the mechanics of Newton, though it naturally gives results almost identical to the latter when the bodies being considered are large enough to have few wave aspects.

Quantum mechanics is necessary to describe processes involving particles at very high energy, and is certainly essential to discuss the very early stages of the universe, as in the hadron era and before. However, it has not yet proved possible to develop a quantum mechanical formulation of particles interacting through gravity at very high energies such as before the 'time' -43. There are many technical difficulties to be overcome before the marriage of general relativity and quantum mechanics can be consummated. Some of them are extremely difficult indeed, and it has been suggested that they may be insuperable. However, recent developments in the theory of radioactivity lends credence to the suggestion that this will not be the case, and that there may even not be the expected increase in the strength of gravity at the 'time' of -43. If that is so then the hadron era may well persist at all earlier times. As 'time' is rolled ever further back the universe may present an ever-similar aspect. There would always be activity as heavier and heavier hadrons became responsible for the structure of the universe. In such a picture one might even be able to say: in the beginning there was no beginning. But it would always get hotter!

There seems to be only one feature which the hadron era can have specifically determined in the universe we inhabit at present, and this is the distribution and sizes of the galaxies. As Professor McCrea has already said, it has proved impossible to understand their formation from initial fluctuations at any later stage in the universe. It may be that earlier irregularities will be able to grow sufficiently to produce the galactic sizes we can observe at present. That is one of the challenging postulates that cosmology will have to investigate.

Penrose has already suggested in Chapter 9 that we could discuss the behaviour of the final state of matter in a black hole or in the collapsing stage of the universe along the lines we have followed for the beginning of the universe. As material becomes ever more and more compressed it will pass successively through various stages similar to those inside a neutron star at ever-greater density, till finally it will enter the hadron era. It will travel through that till it is compressed into a region smaller than the size of the lightest hadrons, which will be broken into heavier companions. Again it will be natural to introduce 'time', the logarithm of time, to describe the subsequent evolution, although we must remember that this 'time' will be different from that experienced by an observer outside the black hole. In the quantised hadron era, when 'time' is less than -23, there may be, as for the very early stages of the big bang, no change of scenario. Each further step of 'time' may appear to be like the preceding step, though now with everything shrunk a little further and all constituents having higher masses. In this quantised situation there will be no end if the correct measure of 'time' is used.

The problems presented by the neutron star, the beginning or end of the universe, or the final state of matter inside a black hole cause present scientific understanding to be stretched to the extreme. But the difficulty raised by trying to apply quantum mechanics to the whole universe is of such a fundamental character as to cause the whole theory to be questioned. To see where this difficulty lies we have to go back to the basic concept of quantum mechanics, that the wave function of a system contains a description of the results of experiments on many copies of the same system.

But how can we apply this idea to the whole universe? There can be only one universe, not many copies of it, and so this probability interpretation would seem to break down. It would appear that a quantum mechanical description would collapse if we applied it to an ever-increasing portion of the whole universe, however small that universe might become. This is very disturbing to a scientist. The basis of scientific method is to explain the properties of an object by means of those of its constituents. Here we find that a description thought valid for the constituents cannot be extended to account for the properties of the whole.

There are two ways of dealing with this dilemma. One is to admit that quantum mechanics is itself only an approximation to the truth valid over small distances, and it must be replaced by a more appropriate theory when applied to the universe as a whole. This new theory must also be able to explain quantum mechanics in the small. We have no hint as to what this theory might be like, especially because it has proved so difficult to avoid the probabilistic ideas of quantum mechanics completely, yet derive them from an underlying deterministic theory.

An alternative to this, only possible for an infinite universe, is to require that quantum mechanics only be valid for every finite volume. In such a world there might always be the possibility of having a large number of copies of a given region, especially in an isotropic homogeneous universe. So quantum mechanics would not have to be destroyed in this model, only accepted as not applicable to the infinite. But the conceptual difficulties that arise when trying to imagine how the universe could be infinite in extent while still in the throes of the big bang quickly become insurmountable. The argument can only be continued in terms of higher mathematics and, believe me, argument will be a particularly apt word to use over the exciting decades to come!

QUESTIONS
WITHOUT
ANSWERS
Professor
John
Taylor

In the introduction to Chapter 1, I referred to the comparison between cosmology and theology. By now it will be clear that these twin edifices of intellectual endeavour stand on the common ground of experience but do not interact at all. This far from satisfactory situation places the cosmologist in a dilemma. As a scientist, is he morally justified in adopting non-scientific categories of thought in order to entertain theological concepts—concepts he often needs to give himself some comfort in an otherwise alien universe? Professor John Taylor is one who is seeking to describe the universe entirely in terms of one aspect of it—the physical. He regards progress along this path as having been so rapid and successful that it leaves very little room for a theological view of the universe at all. And he maintains this approach even in the face of what he describes as 'the unanswerable questions' about the cosmos. L. H. J.

The really important questions about the cosmos have not been answered or even asked, in this book, full though it is of discussions by learned cosmologists. This avoidance is basically because such questions are so difficult to answer as to be a real embarrassment, so we shrug them off by the traditional reply that science is inherently unable to cope with them. How else can we respond to such posers as: if the universe was initially composed of elementary particles where did they come from? What caused the source of the primeval particles to be there in any case? Behind all the machinations of matter, we are here, somehow by a miracle. Is this not evidence of some divine plan? Are we not limited in what we see around us by our own ways of thought; if we had different brains could we not somehow have a different logic and see a very different universe?

In Chapter 11 I tried to demonstrate how it was possible to apply the scientific understanding we have built up so far on earth to the far-flung reaches of the heavens. I painted there a tentative picture of the behaviour of matter under the most extreme conditions. But to answer the impossible questions I have listed above, it would seem necessary to jettison the scientific approach, if only because the bases of these questions do not appear to involve processes observable from our limited vantage point here on earth. This makes me extremely suspicious. If we accept that the scientific approach has been highly successful in extending our understanding of the universe about us, from the application of technology to our daily comforts right out to the dynamics of the solar system and beyond, then why should it fail for questions of this kind? Could it be that the questions themselves are somehow invalid?

Chapter 12

In case you think I am trying to dodge the issue, I must point out that a great deal of scientific evidence *is* relevant to the unanswerable questions, some of it so weighty as to begin to indicate the nature and even the detail of the resolution of these problems. Deep-rooted beliefs can still hold sway to prejudge the issues involved, but I want to show here how one has to close ones eyes ever closer if one wishes to take notice of the evidence, because it is piling up.

One of the most important problems which science has been tackling with ever greater success is that of the nature of life itself. For it is that phenomenon, of which we feel ourselves to be the greatest exponents, that is thought by so many people to indicate a divine purpose. The spontaneous evolution of living organisms from the primordial soup appears to be so improbable, even from a scientific stand-point, that it is easier to believe that life was created with the aid of some mastermind.

There are certainly many steps in the long journey from unfeeling matter to ourselves, and it does indeed seem un-likely, on the face of it, that they were all taken, one after the other, without a guiding hand. Yet as the details of the various steps are slowly being unravelled their automatic character is becoming clear.

The first step to be understood is that of the creation of the earth itself. For many years it had been thought that our globe had been formed by the cataclysmic near-approach of another star to our own sun. The close encounter and the ensuing tidal forces caused material to be drawn out of the sun and this material later formed the planetary system of which we are a member. Such a close collision can be calculated as being so extraordinarily rare as to be well-nigh providential; we would very likely be the only planetary system in the galaxy. But this catastrophic theory of planetary formation is no longer accepted, for various reasons. One is that calculations have shown that fragments drawn out of the sun in this fashion would have soon fallen back into it; another is that the mechanism gives no explanation of the fact that the planets have fifty times more spin around the sun's centre than the sun has itself.

An alternative mechanism, now regarded as the most likely mode of planetary formation, is the condensation of the sun and the planets from a common cloud of gas. This allows a far higher proportion of stars to possess planetary systems. As many as one star out of four could have such small companions, as recent observations have shown. Of course, it is very difficult to observe planets of distant stars directly due to their minute size and the fact that they only

shine by star-light reflected from their surface. But their
gravitational attraction on their parent star can disturb its
motion somewhat, and careful analysis of nearby stars has
indicated the existence of planets round three of them,
Barnard's star, the brighter component of the double star
61 Cygni, and another nearby star, number 21185 of Lalande's
catalogue, all within twelve light years from us. Since there
are about seventeen stars up to about that distance away,
this gives the figure quoted above of one star in four as
possessing a planetary system. This figure is supported by
the fact that a high proportion of stars have visible companions. The remainder very likely have companions
which were too small to form fully-fledged stars; the gas
cloud instead formed planets, some of which are now becoming visible on careful investigation.

In order for intelligent life to develop on a planet, many
other features must be present beyond the planet's mere
existence. From fossilised evidence here on earth it is clear
that life has taken about 3,000 million years to evolve to
ourselves. The necessary conditions for this development to
occur are very critical: for example it cannot have been
too hot or too cold, nor could there be too great variations
in temperature over the day or year. This reasonable
constancy of temperature is essential for living organisms,
such as ourselves, constructed and running on metabolic
processes which are based on the solvent properties of
water. We could not have evolved if the temperature of our
environment had been too close to freezing or boiling point
for a considerable portion of the day.

These considerations rule out a large proportion of the
planetary systems around binary or multiple star systems, as
life-supporting planets. About one half of all stars may be
coupled to others in this way; of the total of fifty-five stars
visible within fifteen light years only thirty-one are single
stars. They also indicate that even round a single star only
a small proportion of the planets will have a suitable temperature range to allow life, as we know it, to develop. In our own
planetary system we only expect Venus and Mars to be
reasonable candidates. Mercury is too close to the sun to be
cool enough and Jupiter, Saturn, Uranus, Neptune and
Pluto are too distant to be hot enough.

Our discussion may be unnecessarily restricted if we
base it only on life as we know it on earth. Serious sug-
gestions have been put forward that life need not, for instance, use water as the universal solvent but ammonia
instead. This would allow for the development of life in a
far wider range of conditions, in particular on a planet such

159

as Saturn which apparently has an atmosphere composed of hydrogen, methane and ammonia.

If we trace some of the further steps involved in the development of life here on earth it is known by experiment that the non-living materials very likely present in the oceans over 3,000 million years ago could have produced very primitive living material by the action of sunlight or electric storms. In the now classic experiment in 1953, Stanley Miller at the University of California in Los Angeles subjected a mixture of water, hydrogen, carbon dioxide, methane and ammonia to electric discharge and obtained amino acids and other components of living matter in surprising quantities. These organic compounds can be utilised to build up larger molecules, and while the most suitable environment for this is yet unknown there is no doubt that such conditions do exist.

It would be illogical to expect it to be easy to duplicate in a laboratory processes which took up to 1,000 million years to achieve in the natural condition on earth. Many of the further steps leading to the production of the amazingly specific enzymes which catalyse only one or two out of possibly thousands of reactions, or of the nucleic acids which carry the genetic message of the cell for its reduplication or function, are mysterious and will undoubtedly remain so for some period ahead.

Yet the general mechanism of this specificity and of the constancy of the genetic message are now understood; the selectivity of enzymes is considered to be due to the particular three-dimensional 'coiled' structure assumed by the linear chain of amino acids of which the enzyme is built. This stereo-specificity, as it is called, of enzyme catalysis works exactly as a key does in fitting into a lock: it has to have the correct shape. The preservation of the genetic structure is assured by the Watson-Crick spiral helical model of DNA and RNA, with its code determined by a suitable order of linkages between the two strands of the helix.

The basic aspect of evolution, the selection by the environment of the most suitable of all available genetic types, has been finally placed on a firm basis—by this understanding of the biological process at a molecular level. Genetic mutations caused by incoming cosmic radiation or chemical effects allows a constant variability in genotype to have been available upon which the environment can operate to select the most suitable. Such an evolutionary process undoubtedly occurred at the earlier stage of molecular evolution. The environment would have affected the completely random linking of various amino acids into proteins so as to allow only a very selected range of them, these being the only ones able to exist under such conditions. The difficulty of the improbability of any particular protein being formed if billions upon billions of various proteins are equally likely to be created is only apparent, and arises only because of neglect of the surrounding environment. This environment is continually and subtly shaping the molecules so that only the 'right' ones are formed.

There is a further problem: that even the simplest living organisms apparently violate one of the basic laws of science —the second law of thermodynamics, that chaos must always increase. But this is also a fallacy. For though life can only grow by increasing its own organisation and decreasing its chaos it does so at the expense of its environment; it spreads chaos around it, as we can see only too well in the effect of mankind on his surroundings. Recent experiments have even shown how localised highly organised states can arise spontaneously in non-living systems, though they can only persist for a finite time. But then death is a property of all living systems; return to the state of general chaos must always occur sooner or later.

The process of evolution has led, somewhere between the viruses and ourselves, to the evolution of mind. One of the great enigmas of modern science is to understand how this remarkable facet of living matter actually arose. We do not doubt that sooner or later this puzzle will be solved for the lower animals. Their brains involve comparatively few nerve cells, and pose only a reasonably severe problem to scientific analysis. It is as we climb higher up the evolutionary tree that the question of the mind rapidly becomes intractable. Yet despite a persistent belief to the contrary there is no reason to suspect that the mammalian mind, and particularly that of man, is qualitatively any different from that of lower animals. Behaviour is controlled by a brain constructed in the same general fashion and out of the same building blocks for all animals. Nature is always economical in the use of material; it is only the pressure of evolution which has led to brains composed like ours, of 10 billion nerve cells, with a correspondingly high complexity of connections.

In support of this reductionist approach is the success of brain research in controlling and correcting all aspects of mental states by suitable physical modification of the brain. There has not been found any hint of an escape from the conclusion that it is the brain which is all powerful, and not the mind. This is in complete contradiction to the popular feeling that our mental world is really the most basic, yet it is a result for which the evidence is now very strong. That is not to say that the mind plays no role, but that it is a role determined completely by the physical processes proceeding in the brain.

From what I have said above there is no reason to doubt that brains and minds can and will develop on other planets which are hospitable to life. Such thinking life could in some cases have advanced far beyond our own understanding of the world around us. Civilisations in our galaxy will probably have far outstripped our own in technological development. For some years now, various scientists have been listening for tell-tale signals from outer space (Frank Drake in Project Ozma, for example, where an 85-foot radio telescope listened to likely sources in the nearby stars of T Ceti and E Eridani).

The search is concentrated on radio emission using the 21-centimetre line of hydrogen. This is a frequency which would be recognised as universal to all sufficiently intelligent communities due to its intrinsically molecular origin. It is also particularly useful since it suffers least interference both from radiation from the earth's atmosphere at the high frequency end of the spectrum and from the galaxy at the low frequency end.

The three-months' duration of the Project Ozma search in 1960 produced no results, but since then calculations of the expected power which alien civilisations could radiate, compared with the sensitivity of the receiver Drake used, have indicated that this null result was to be expected, even if highly intelligent alien life did exist. Much higher sensitivity is needed to detect possible messages from afar, as well as a great deal of luck so that such signals would be beamed in our direction while we were looking for them. All in all, while we have no direct evidence of alien civilisations in the galaxy they are expected to be there. Informed guesses put the figure at at least one million—or one in the 10,000 stars within a hundred light years of the sun.

If there is intelligent life elsewhere it need not be at all like ourselves, and may very likely observe completely different features of the universe from those which we observe, at least initially. This selection will depend heavily on the main features of the environment in which the intelligence develops. With an atmosphere such as ours in which light gives a very clear indication of the surroundings undoubtedly light-sensitive receptors would play a strong role. But the nature of the receptors used to determine distant food or threats, for example, may be far different from our own; electric currents could be used in a dense conducting atmosphere, such as water or ammonia. Electromagnetic radiation of very different wavelengths than present on earth might be of great importance as a means of sensing in a dense atmosphere; for example a radar-type of detection could be used by beings in such an environment. A high-gravity planetary surface would be conducive to highly muscled, quick-reacting creatures with the ability to counter-act the strong attraction of their planet.

The nature of the environment will strongly determine those of its features which intelligence will first try to under-stand and control. Thus when electromagnetic radiation is an important mode for sensing then this will soon be in-vestigated in the development of civilisation. Gravity itself will pose the main puzzle on a high-gravity planet, and other intelligences may be far beyond us in an understanding of this phenomenon. Evidently the development of science in such communities may be along very different lines from here on earth. However, we would expect that the same general picture of the universe would ultimately emerge for alien cultures that has for us. After all, we have extended our natural senses by artificial means to span the entire range of the electromagnetic spectrum, and are even now attempting to probe that of gravity. We would expect all intelligence eventually to arrive at the same general picture of the world as they developed special instruments to fill the gaps in their own powers of observation.

It would seem that the logic of the world will be the same, whatever the alien civilisation observing it. If it were pos-sible to conceive of intelligence emerging at an atomic level its resulting world-view would admittedly be logically different from ours. The probabilistic nature of events would rapidly be evident to such creatures, playing an unconscious role in their being until it was clarified as an objective feature of the universe. The science, if such it could be called, of such a civilisation would be founded on lack of certainty in all things. However, it is not clear that science could in fact grow from such an unsure foundation. Even communication would be a problem, since its essence is to transmit and receive information with the least possible amount of distortion, and distortion could be excessive at the atomic level. Assuming such problems were resolved it would still be expected that extension of observation of the universe to events over a much larger scale would result in the deterministic world-picture we know and have most direct experience of on earth. Only if the world at that level could be manipulated would atomic-sized intelligence be able to develop any technology for effectively controlling their world. Evidently such minute intelligences would have many difficult steps to take before they could become a viable intelligent community. But when they had taken these steps they will have followed along the path of science as we know it, with its emphasis on objectivity and consensus. For without these twin foundations, any understanding of the world would only be built on shifting sand, repeatedly to sink and be buried by lack of predictive power and of the parallel understanding it produces.

It would seem, then, that our view of the cosmos, starting as it has done from a limited spectrum of the observable world, but now broadening to include a much greater range of phenomena, is one which would be reached by any intelligent civilisation in this or any other galaxy. Some cultures will have developed further in certain aspects, others less, while undoubtedly certain of them will have gone very much further than us in all the scientific disciplines. But they will all have to tread the straight and narrow path of science, otherwise their conjectures and ideas about their world will only travel down the primrose path to that everlasting bonfire which steadily consumes ill-founded hopes and beliefs.

For example, let us suppose, contrary to the strong indications I stated earlier, that mind is in fact a separate entity from the brain, so that the two can exist independently. It is difficult to conceive how this is possible in any detail, and so hard to put such an idea to any worthwhile experimental test. But if it were found to be true by some alien culture it would lead to the following very curious state of affairs. Matter has not always existed in a form to which mind could hook up to. At the early stages of the big bang, especially in the first minutes, there would be absolutely no way in which mind could use matter, due to the enormous physical forces and energies that were then universally predominant. It seems, at least from our own experience on earth, that only when the delicate tracery of excitable nerve cells and their connections can be allowed to exist peacefully over periods of time that matter is in a suitable state for mind to relate to it. For us this only occurred at least 7 billion years after the big bang. This liaison may have been feasible far earlier on other planets elsewhere in the cosmos, but there is no way of envisaging such a hook-up when temperatures were much above 10,000 degrees and all matter was at least in liquid form if not in a far more chaotic state. This means that for at least the first million years of the universe mind and matter were completely separate; they very likely were for at least a thousand times longer than that.

But we are left with the disturbing problem as to where

mind was during this long period. Indeed it could have had no energy, since otherwise it would have been affected by the interactions between matter involving gravity. In particular no physical organisation could have been possessed by the universal mind extant in the first few seconds, since otherwise it would have been chewed out of existence by the all-dominant radiation of various forms bringing any localisations of energy to the same completely chaotic level.

Mind must have been purely non-physical to survive such a holocaust. The question then is how it ever became able to relate to the developing local organisations of matter at all. It seems inconceivable that mind could ever have built a bridge to the physical world. It is as if one had to build a bridge across a river, one side of which was composed completely of unfathomable quicksand. How could a firm foundation ever be put down on that side which could support the bridge? If one tried to work from the other side of the river, even using the most powerful foundation possible, the bridge would still fall unless all the weight was carried on the solid side. But then it would not really be a bridge at all; it would be an extension of the road from the solid physical side to the uncertain shifting one.

That is the difficulty with a purely non-physical mind trying to relate to organised matter. There seems no foundation in the mental side which can ever direct or control physical energy. Non-energy has suddenly to become energy, and it seems impossible for it to do so.

From these remarks it follows that it would be almost impossible for any intelligent civilisation to conceive of the mind other than as a very useful adjunct of the physical brain. On this view the universe then acquires a beautiful unity: it is composed purely of energy represented in various forms, all transmutable into each other under certain conditions. The mental world has been dissolved and become one with the physical universe. Any search for divine purpose or a creator of the world can, for the scientist at least, only be at the level of energy.

Purpose, as we normally interpret it, can have no meaning in such a unified cosmos, now that our very self-centred views of the universe have been discarded. For how can energy have a purpose? It can have power, flow and direction in space. Indeed one can say that energy creates space by its very presence. Without energy space cannot even be experienced. Nor do we expect time to exist without energy, since there would then be no possibility of change, no way of noticing the flow of events.

In the energetic universe the answers to the unanswerable

questions posed at the beginning of the chapter become restricted but beautifully coherent. If there is no divine purpose at work, 'thinking' through the laws of the world which we are slowly stumbling upon by our laborious scientific method, then there can be no 'act' of creation at all, as such. We have already dimly glimpsed, in the previous chapter, a possible beginning for the universe in which there was no beginning. As the clock was rolled ever back, closer and closer to the point of time we initially regarded as the first point of existence, there appeared to be ever-greater activity. To account for this we introduced a different time, there called 'time', which took proper account of these ever-present interactions between elementary particles.

The problem of the creation of the world is seen to be incorrectly posed. We are in a phase of the development of the universe in which time, the measure of activity which we most immediately experienced, is quite suitable. But we cannot use this same measure to extrapolate back to the very earliest stages; the more correct 'time' has to replace it. And 'time', a measure of the activity in the cosmos, had no beginning: it was 'always' there.

There is still one impossible question which I did not list among those at the beginning of the chapter, basically for the same reason as that I gave for why the unanswerable questions are usually avoided by scientists. The problem is that even if the universe is energy, why is it here at all? The form of the question requires some idea of purpose in order to answer it. I have claimed above that purpose is only a dream in our minds, and is not visible in the universe.

The only answer I can give to this ultimate of impossible problems is that there is no reason at all, no purpose, nothing. The universe just is, as energy. The laws which govern the interactions between localised aspects of energy, laws which we are slowly sensing, give form and structure to it. These rules also just are, without reason or purpose. In other words the basic facts of life about the universe are just that, facts, and no more.

It is only if we recognise ourselves as one manifestation of energy among many in a unified universe that we can begin to come to terms with this most difficult feature of all we will ever have to face. And it is by the endeavours of cosmologists and other scientists that the nature of the unity can become ever plainer, so that eventually we may be able to recognise in ever-greater detail the universe for what it truly is: energy made manifest.

INDEX